Palgrave Advances in Bioeconomy: Economics and Policies

Series Editor

Justus Wesseler, Agricultural Economics and Rural Policy Group, Wageningen University, Wageningen, Gelderland, The Netherlands

More information about this series at
http://www.palgrave.com/gp/series/16141

Rita Vilkė · Dalia Vidickienė ·
Živilė Gedminaitė-Raudonė ·
Vitalija Simonaitytė · Erika Ribašauskienė

Rural Economic Developments and Social Movements

A New Paradigm

palgrave
macmillan

Rita Vilkė
Lithuanian Centre for Social Sciences
Vilnius, Lithuania

Živilė Gedminaitė-Raudonė
Lithuanian Centre for Social Sciences
Vilnius, Lithuania

Erika Ribašauskienė
Lithuanian Centre for Social Sciences
Vilnius, Lithuania

Dalia Vidickienė
Lithuanian Centre for Social Sciences
Vilnius, Lithuania

Vitalija Simonaitytė
Lithuanian Centre for Social Sciences
Vilnius, Lithuania

ISSN 2524-5848 ISSN 2524-5856 (electronic)
Palgrave Advances in Bioeconomy: Economics and Policies
ISBN 978-3-030-71982-1 ISBN 978-3-030-71983-8 (eBook)
https://doi.org/10.1007/978-3-030-71983-8

This Palgrave Macmillan imprint is published by the registered company Springer Nature Switzerland AG
The registered company address is: Gewerbestrasse 11, 6330 Cham, Switzerland

Praise for *Rural Economic Developments and Social Movements*

"Revolutions are born from the bottom up, growing in a small group of people to explode and change reality. It is the people, their charisma and ideas that change the established paradigms of development, and the role of good researchers is to capture these processes, name and document them, and lead us to consider their effects in the future. This is the case with this book, as it proves that social movements, which have been the subject of research for decades, are still an important part of how economies and societies change and progress. This very timely book is written from the perspective of experienced researchers and sensitive observers of rural change, and it is worth reading because it describes the processes that have brought about the green transformation and are now changing not only rural areas but the entire global economy."

—Dr. Paweł Chmieliński, *President of the European Rural Development Network and Professor at the Institute of Rural and Agricultural Development, Polish Academy of Sciences*

"This monograph is a timely and significant call to take a new look at numbers of already matured and ongoing innovations in rural areas around the world. Authors of the book masterfully use the grounding of the social movements' theory to the ongoing processes in rural areas, thus proposing a brand-new consideration of the rural paradigm shift. To disclose the reasoning behind the social movements, the monograph takes focus on actors of change and gets deeper into the context of ongoing organized changes in rural areas, distinguishing between the industrial paradigm grounded rural social movements and new post-industrial change-focused rural social movements, putting them into the general context of European Union Common Agricultural Policy (CAP) and European Green Deal (EGD) principles. This monograph is based on a holistic approach. Systematic, evolutionary, and actor methodologies were used. Finally, in-depth illustrations of ongoing processes using multiple case studies of rural social movements significantly add to the existing body of knowledge in this field of science."

—Prof. Dr. Maria Nijnik, *Principal Scientist of The James Hutton Institute, UK, Coordinator of H2020 SIMRA project "Social Innovation in marginalised rural areas"*

"The remarkable book, which applying a rich methodological apparatus of qualitative research step by step reveals the role of individual and territorial social capital for the post-industrial transformation of rural territories, it offers more questions than answers. The holistic approach used by authors simply does not allow to formulate unambiguous answers in the context of the investigated causes of genesis and impacts of (new) social movements on innovative, sustainable, and inclusive rural development. There does not exist only one generally valid and accepted model of explanation and assessment for changes induced by spatial redistribution of population, uneven diffusion of new types of knowledge and practical involvement of local actors with different interests and interpretations of rural development. Through tacit and/or explicitly formulated questions, the authors 'play an exciting game' that will delight anyone interested in a deeper understanding of the ongoing transformation of the European countryside."

—Dr. Vladimír Székely, *senior researcher of Institute of Geography, Slovak Academy of Sciences*

Contents

About the Authors

Dr. Rita Vilkė, Ph.D. in Social sciences (Management and Administration) is a principal researcher at the Lithuanian Centre for Social Sciences, Institute of Economics and Rural Development. Her research interests are recently focused on the broad issue of corporate social responsibility in economically viable and sustainable agriculture and rural development, innovative approaches to collaboration and governance, value chains and modern management systems for agribusiness, as well as its impacts on regional and global sustainable development. Rita Vilkė is an author and co-author of more than fifty scholarly articles, chapters of books, the reviewer/board/advisory board member of Emerald book series, referred international scientific journals and conferences. She holds expertise in the implementation of international scientific research and development projects. She participates in internships and exchange programmes (UK, Germany, Netherlands, Finland), international scientific conferences with presentations, moderate sections. She is a member of the global Social Responsibility Research Network, Eurasia Business and Economics Society, European Association of Environmental and Resource Economists, Lithuanian Young Scientists Organization.

Dr. Dalia Vidickienė, Ph.D. in Social sciences (Economics) is chief researcher at the Lithuanian Centre for Social Sciences, Institute of Economics and Rural Development with forty years of research experience in economics and management. Her main interest centres on rural development and regional innovation policy, strategic management and business models. She has a long time experience in the management and coordination of international research and development projects of different EU programmes (FP6, INTERREG, PHARE, Leonardo da Vinci). Currently, she is focusing on research topics dealing with paradigm innovations in rural and regional development caused by the shift into post-industrial servitizated and knowledge-based economic system.

Živilė Gedminaitė-Raudonė, Ph.D. in Social sciences (Economics) is a principal researcher at the Lithuanian Centre for Social Sciences, Institute of Economics and Rural Development. The main research focus is sustainable regional and rural development, climate change, circular economy, innovation ecosystems, and transformative tourism, among others. Zivile Gedminaite-Raudone is an author and co-author of more than forty scholarly articles, chapters of books. She has rich experience in management and coordination of many international research and development projects of different international, the EU and national programmes focusing on new knowledge creation using co-creation and cooperation principles. She actively participates in internships and exchange programmes (Italy, Belgium, Ireland), international scientific conferences with presentations, moderate sections. She is a member of the global Social Responsibility Research Network, European Rural Development Network, Eurasia Business and Economics Society.

Vitalija Simonaitytė, Ph.D. in Social sciences (Political sciences) is a researcher at the Lithuanian Centre for Social Sciences, Institute of Economics and Rural Development. Her recent research focus is sustainable regional and rural development policy, interest groups, and political parties, local and regional actors. She is an author and co-author of publications in various scientific journals and research studies, also she participates in international scientific conferences (ECPR, CEPSA) and international research projects (Comparative Interest Groups Survey, CIGS; European Social Survey, ESS). She has experience in management and coordination of rural policy planning and implementation. Currently, she is implementing the EU common agricultural policy planning at a national level in the field of agricultural knowledge and innovation systems (AKIS).

Erika Ribašauskienė Master in Social sciences (Management) is a researcher at Lithuanian Centre for Social Sciences, Institute of

Economics and Rural Development. Her research recent focus is sustainable regional and rural development practices, local and regional stakeholders, rural crafts and business models. She is an author and co-author of publications in various scientific journals, research studies and books. She has experience in management and coordination of rural policy planning and implementation in many areas of rural development. Currently, she is implementing the EU common agricultural policy planning at a national level in the fields of social welfare and economic development in rural regions.

Abbreviations

AIC	Agricultural Industries Confederation
ARC	Agricultural and Rural Convention
CAP	Common Agricultural Policy
CELCAA	European Liaison Committee for Agricultural and Agri-Food Trade
CLLD	Community-Led Local Development
COGECA	General Committee for Agricultural Cooperation in the European Union
COPA	Committee of Professional Agricultural Organizations
COPA-COGECA	Joint Organization of Committee of Professional Agricultural Organizations and General Committee for Agricultural Cooperation in the European Union
EC	European Commission
ECOVAST	European Council for the Village and Small Town
EGD	European Green Deal
EIP	European Innovations Partnerships
ELARD	European LEADER Association for Rural Development
ELO	European Landowners' Organization
ENRD	European Network for Rural Development
ERCA	European Rural Community Alliance
ERDN	European Rural Development Network
ERP	European Rural
ESIN	European Small Islands Federation
EU	European Union
FAO	Food and Agriculture Organization of the United Nations
GEN	The Global Ecovillage Network

GMO	Genetically Modified Organisms
ITMN	Indigenous Terra Madre Network
LAG	Local Actions Group
LAG Network	Local Action Group Network
LEADER	(in French: 'Liaison Entre Actions de Développement de l'Économie Rurale'; in English: 'Links between activities for the development of rural economy')—EU support scheme for the rural economy through local leaders.
LFFU	Lithuanian Family Farmers' Union
LFU	Lithuanian Farmers' Union
LRP	Lithuanian Rural Parliament
LURC	Lithuanian Union of Rural Communities
LVC	La Via Campesina International Peasants' Movement
NGO	Non-governmental Organization
NRN	National Rural Network
NSM	New Social Movement
OECD	Organization for Economic Co-operation and Development
RDP	Rural Development Programme
SDG	United Nations Sustainable Development Goals
SFYN	Slow Food Youth Network
UIPI	International Union of Property Owners
UNFCCC	United Nations Framework Convention on the Climate Change
URGENCI	Community-Supported Agriculture and Forum Synergies platform
WFO	World Farmers' Organization
WTO	World Trade Organization

List of Figures

List of Tables

CHAPTER 1

Introduction

Rita Vilkė

Today's society calls for a new explanation of recent transformations around the world. The new innovative, sustainable and inclusive rural development paradigm demands original qualitative dimensions of researching, exploring and explaining rural socio-economic development and rural policy transformations in light of overall modern society's social change. The new rural development paradigm stresses the importance of bottom-up policymaking, self-organization, creative use of knowledge in rural areas, and other innovative success principles. In this book, the ongoing transformations are aligned with a new look on social movements' theories and approaches from a paradigm innovations point of view, which help disclose, explore and explain the already ongoing rural paradigm shift in post-industrial society's development.

Even though the area of social research, devoted to social movement phenomena, is already known for more than a century, it performs significant reorganization from a scientific point of view, especially in the last few decades. Throughout the century well-examined, explored and explained in scientific research, collective behaviour in a form of social movements start demonstrating the proliferation across the social sciences disciplines (Roggeband & Klandermans, 2017; Waddock, 2017). Social movements as an object of scientific debates had already entered the new era of post-industrial economy and knowledge society, with original discourses and implications. A quick overview of the interest in the field demonstrates

R. Vilkė et al., *Rural Economic Developments and Social Movements*, Palgrave Advances in Bioeconomy: Economics and Policies,
https://doi.org/10.1007/978-3-030-71983-8_1

the increase, as well as the contextual proliferation of social movement phenomena in the past three decades. Between 1990 and 2000 scientific investigations concerning new discourses of social movements counted 837 thousand scientific publications, in 2000–2010 more than doubled to 1.740 thousand, and in recent decade they keep slight growth till 1.810 thousand investigations in the field (Google Scholar Keyword Search Engine, 2020).

Particular attention in scientific discussions is devoted to the new social movements (NSMs), especially—in their so-titled 'explosive power' concerning the quick mobilization of masses and resources to make the social change (Roggeband & Klandermans, 2017; Waddock, 2017, etc.). NSM roots arrive from so-called classical, old or industrial social movements, which are a well-known form of protests going back to the eighteenth century (Castells, 2015; Crozat et al., 1997; Larana et al., 2009; Lofland, 2017; Tilly et al., 2019). Initially, social movements were understood as a collective action of similarly minded people against existing arrangements of power and institutions as well as institutional actors in the particular issue (Crozat et al., 1997; Tilly et al., 2019).

The mentioned classical, also known as old social movements, use to hold five basic characteristics (Lofland, 2017): first, they were rushed by independently and newly founded or rejuvenated protests or violent organizations; second, they should hold a rapid rise of participants as well as numbers of planned and unplanned protests and violent actions; third, they should necessarily be encouraged by mass opinion in a particular issue; fourth, represented mass opinion should be addressed to the agents of central institutions; and finally, they should respond to emerging changes in either macrostructure or central institutions. In other words, social protests were grounded in class conflict and confrontation for the justice and human social being issues most often concerning labour, political decisions, gender, equality of rights, regionalism, etc. (e.g., Larana et al., 2009; Tilly et al., 2019). In turn, especially when times of industrialized economies emerged, social movements gained more institutionalized forms of representation (Crozat et al., 1997), so they start appearing in scientific investigations concerning the industrial paradigm and industrialism-grounded social movements (Coles, 2004; Hess, 2016; Kivisto, 1984; Sen & Lee, 2015; Touraine, 1985).

So-called NSMs in Western Europe start being examined by scientists in the late 1960s. In contrast to the old class conflict-based social movements, NSMs appeared with new power by stressing reconciliation of interests of different groups of society aiming to obtain the maximum positive synergistic effect in a particular issue. The main observed distinctive feature of NSMs was a shift from conflict character between society and particular institutions to the promotive role of particular prospective value-based belief to become an organized acting structure, moreover—driving philosophy of particular values-based groups of individuals, composing new local communities (e.g. which later appeared in a form of particular ecovillages); or even broader—become bord, moreover—continent-crossing powerful international society (e.g. La Via Campesina peasantry movement). Initially, NSMs emerged as an organized bottom-up reflection to objective societal problems such as environmental degradation and safety, women exploitation, healthy food, etc., and did not strive for political participation in institutions (Inglehart, 1990). From the other side, the power gathered through the circulation of particular ideas through already acting networks did mobilize public opinion and therefore affected political decisions due to relevant issues.

NSMs received great attention from the scientific community (e.g. Brand, 1990; Buechler, 1995; Cohen, 1983; Inglehart, 1990; Melucci, 1980, 1993; Touraine, 1971, 2002; Wieviorka, 2005, etc.) and became well-documented phenomena throughout the last half of a century. In the light of the twenty-first-century NSMs had already been titled post-industrial social movements (Clark, 2018; Doherty & Doyle, 2006; Melucci, 1993; Sutton & Vertigans, 2006). Lately, the ongoing modern transformations in rural regions had been allocated next to the new post-industrial social movements, by highlighting, that the emergence of NSMs greatly affected the new reconstruction in the rural development paradigm (Deere & Royce, 2009; Rosset & Martínez-Torres, 2012, Rosset et al., 2013; Woods, 2003 etc.). However, any comprehensive scientific studies that would be driven to holistically disclose the rural development paradigm shift in the light of social movements theory had not been observed in scientific literature yet.

This monograph is devoted to go through relevant scientific debate and get deep into the actual practice of the rural development paradigm shift, accelerated by a new form of consolidated power of social change—a new generation of social movements that act as paradigm innovators and active co-creators of rural development policy. To disclose the actual reasoning for an ongoing rural paradigm shift due to the social movements, the monograph takes focus on actors of change in rural areas, including empowered and free stakeholders (farmers and their organized structures, agricultural companies, cooperatives, ecovillages, local rural communities and other non-governmental organizations) and gets deeper into the context of ongoing organized changes in rural regions by fields in which actors make a change, distinguishing between the industrial paradigm grounded rural social movements and new post-industrial change-focused rural social movements, putting them into the general context of European Union Common Agricultural Policy (CAP) and European Green Deal (EGD) principles.

In contrast to other research in the field, this monograph is based on a holistic approach. Systematic, evolutionary and actor methodologies were used in the monograph. To better disclose the diversity of researchers' attitudes towards the new social movements and rural development paradigm shift, as well as the overall patterns of social change, which comes as an outcome of analysed phenomena, the theoretical pathway of the emergence of a new type of social movements for rural development had been inspected using the systemic methodology and historical approach. Identification of rural development paradigm shift factors at the macro level was done using the evolutionary methodology and distinguishing between industrial and post-industrial development stages of the economic system and society. A new theoretical perspective for understanding how social movements can accelerate post-industrial turn by changing the mental models of customers, entrepreneurs and policymakers had been proposed. The actors' approach helped to highlight the role of the innovator in transforming the industrial villages and farms into a post-industrial countryside using the collective power—rural social movements, who are the producers and users of paradigm innovations. A systematic methodology has been used to understand the determinants of a rural development paradigm shift as part of a larger system, the process of evolution of general economic policy models. Systematic and comparative analysis of rural development history, paradigm innovations, rural and urban actors and their acting environment, determinants of rural social movements' success factors and case studies, using the general outline of the state of the particular type of rural social movement around the world, and in-depth case studies from Lithuania, illustrate how rural social movements corresponds to the determinants of rural development paradigm shift and help to highlight and solving problems of rural areas at the national and regional levels.

The appropriate use of qualitative research methods, such as a case study, helped explain why and how social movements contribute to the rural development paradigm shift in certain cases: first, at the micro-level—social movement-specific, then at the macro level—rural development policy-specific. Besides, the choice of the case study method was prompted by the fact that the review of research on rural social movements revealed that so far the relationship between social movements and rural development paradigm shift had not been a key focus of any systemic scientific research.

The case study approach, according to Yin (2003), is appropriate when certain criteria are met: (a) the main purpose of the study is related to the goal of explaining the object under study by seeking answers to the questions 'why' and 'how'; (b) there is no possibility of manipulating the behaviour of the actors under investigation; (c) contextuality relevant to the study; (d) unclear links between the phenomenon under study and the context. This research outlines the emergence and the state of particular rural social movements around the world aiming to understand the general context. Then the research goes into in-depth multiple case studies in Lithuania to explain why the particular rural social movement emerged in this overall context, how it developed and what role it was played by that movement concerning the phenomena of investigation—the rural development paradigm shift. As all analysed social movements had already been established and their results are visible, it was not possible to influence the behaviour of the research actors and the social movement organization process. Contextuality for rural social movement research is crucial because the context helps to explain what stage of maturity the researched movement has reached and why a certain factor has brought success or failure to the movement development.

To answer the main questions of the case study, i.e. why and how the new post-industrial change-driven rural social movements emerged, and what overall impact they made on the rural development paradigm shift, the auxiliary case study questions were formed following the thoroughly created methodology. An explanatory multiple case study, with carefully selected cases that consistently illustrate the appropriate areas of new rural social movements under paradigm shift in post-industrial rural development, provides an in-depth description of the key success factor mechanism, that would not be possible if using survey or experimental research strategies in this study (Yin, 2003). Cases were selected using the scientific selection strategy as proposed by grounded theory (Glaser &

Strauss, 1967), taking into account the experience from previous investigations in the field when, based on the analysis of the first few investigated cases (Vidickiene, 2013; Vidickiene et al., 2019, etc.), it was decided which elements are appropriate to further inclusion in the research of this monograph. The selection of the most informative cases (Creswell et al., 2007) for this research was done using rich authors' scientific expertise concerning overall rural development in Lithuania and previously implemented research in the field. In total 11 case studies were done by monograph authors: six cases represent industrial paradigm-focused social movements and five cases illustrate post-industrial paradigm and change-focused social movements. Case study data were collected via face-to-face or telephone interviews in July–October, 2020, using a priorly developed tool—an open structured questionnaire (see Table 1.1).

Each question was composed using the analysis results from the life cycle approach of a social movement (see Sect. 2.3). There were distinguished four life cycle stages of a social movement: emergence (What was wrong? What to change?), coalescence (Why to 'go public'?), bureaucratization (How it works?) and decline (How it affected the state of the art?). Each stage included a set of aspects that are relevant to disclose, as suggested by theoretical findings when doing the relevant case study. And then the questions were composed as a structured open-ended framework to collect the relevant data and then develop a case study description. Each interview took from 2 to 3 hours of conversation with a particular movement's leader, who is a driving person of a movement, holding decision-making power. Additional demanded data according to the research framework was collected using document analysis and analysis of various accessible electronic and media sources.

The monograph starts with a comprehensive overview of modern society's development forces that arrived alongside the twenty-first century in a new focus driven social movements and imposed dramatic overwhelming social changes around the world, including the decisive shift in the rural development paradigm. Naturally declined the agricultural sector's role in the rural economy keeps changing the lives of rural communities as well as people's motivation to live in rural areas. Most of these changes are the components of the transition to the post-industrial society's evolution phase, based on knowledge and its creative application. At the same time, these changes resulted in new success factors for qualitative life in rural areas that are fundamentally different from the success

Table 1.1 Case study framework and questionnaire

The life cycle stage of a social movement	*Analysed aspects*	*Questions*
Emergence WHAT WAS WRONG? WHAT TO CHANGE?	History of the emergence of the social movement: occasion, motivation, goals	How did you get the idea to establish this initiative/movement? What problem (s) did you see and want to solve with this initiative/movement? What was your main goal in the beginning and has it changed? What is the date of the establishment of the movement?
Coalescence WHY TO "GO PUBLIC"?	The need, reasons and motivation for growth and expansion of boundaries	Why did you decide to gather like-minded people and share your ideas more widely? How many people are involved? actively? and how much in total (approximately)? What are the main characteristics of members? rural/urban population? small/large farmers? other professions? men/women? younger than 35 years? Is there a change of members in the movement?

(continued)

Table 1.1 (continued)

The life cycle stage of a social movement	*Analysed aspects*	*Questions*
Bureaucratization HOW IT WORKS?	1. Resources 2. Structure 3. Roles and their assignment 4. Identity and location 5. Area(s) of impact 6. Channels	How are the resources mobilized for the activities of the movement (financial, material, human; internal, external)? What resources your movement is lacking? Why? What is the structure of your movement? formalized/informal? centralized/decentralized? How can someone become a member? Is membership open/with restrictions, e.g. recommendations when inviting to become a member or participate in an event/project? Do the members of the movement have their roles, functions or areas of responsibilities? How are they divided? How do you build the identity of the movement? Is your movement geographically localized? Are there any traditional events going on? What channels do you use for internal communication to spread the word about yourself and organize your activities? Do you have an actively used virtual platform? How is it administered, or are there paid employees? Is the movement involved in any international movement?

The life cycle stage of a social movement	*Analysed aspects*	*Questions*
Decline EFFECTS & OUTCOMES	Success Failure Co-optation Repression Mainstream	How would you describe the current stage of your activity? successful? has difficulties/vague? enlargement is still ongoing? static activity? joining global organizations? or other? How do your activities change the environment or those around you? amendment of legal acts has been achieved? redistributed support or subsidies? the harmful activity has been suspended? pilot project (s) for innovative solutions is being implemented? behavioural changes in rural communities, public authorities, agrotechnologies, food consumption habits, etc.?
Maybe you would like to add something from yourself that has not been talked about?		

Source Developed by authors

factors, specific to the previous rural industrialization phase. Social movements in this context arise as drivers of key paradigm innovations in rural development. The monograph originally outlines the three key paradigm innovations and gives a modern explanation of how they occur in ongoing society's transformations.

Further scientific discussion of the book is devoted to the agents of change of rural areas and the political debate regarding 'top-down' (Common agricultural policy, CAP) and 'bottom-up' (European Green Deal, GD) approaches, as well as a common will of the European Union, related to the ongoing rural paradigm changes. The so-called 'just transition' framework encompasses a range of social interventions needed to secure workers' rights and livelihoods when economies are shifting to sustainable production. This has become a recognized element of climate and broader rural development policies. The monograph originally discloses how increased demand for a better quality of life around the world raised many questions in rural regions of how to find the best ways of reconstructing rural areas and how to do this facing the current challenges of CAP and GD. Research discloses how the two existing groups of crucial actors in rural development, i.e. rural population, and, surprisingly—urban population(!) as agents of change play a top-significance role in rural development paradigm shift in a form of social movements.

Finally, the monograph's scientific investigations are illustrated with case studies following the logic of twofold agents in the rural paradigm shift. First agents belong to industrial rural development paradigm-grounded social movements, concerned with the redistribution of property, social, economic power, mass production, added value, risk between farming as livelihood and other sectors. The second ones are agents of change in post-industrial change-focused social movements, concerned with new forms of agriculture, food quality, accessibility, networking among farmers and consumers, new livelihoods and new forms of rural lifestyles.

The concluding part of the monograph puts the reader into summary conclusions and calls to join the original discussion regarding the interconnectivity among social movements and rural paradigm shift, based on previously envisaged investigations in the book.

References

Brand, K. W. (1990). Cyclical aspects of new social movements: Waves of cultural criticism and mobilization cycles of new middle-class radicalism. In *Challenging the political order* (pp. 23–42). Polity Press.

Buechler, S. M. (1995). New social movement theories. *Sociological Quarterly, 36*(3), 441–464.

Castells, M. (2015). *Networks of outrage and hope: Social movements in the Internet age*. Wiley.

Clark, T. N. (2018). *Citizen politics in post-industrial societies*. Routledge.

Cohen, J. L. (1983). Rethinking social movements. *Berkeley Journal of Sociology, 28*, 97–113.

Coles, R. (2004). Moving democracy: Industrial areas foundation social movements and the political arts of listening, traveling, and tabling. *Political Theory, 32*(5), 678–705.

Creswell, J. W., Hanson, W. E., Clark Plano, V. L., & Morales, A. (2007). Qualitative research designs: Selection and implementation. *The Counseling Psychologist, 35*(2), 236–264.

Crozat, M., Hipsher, P. L., Katzenstein, M. F., Keck, M. E., Klandermans, B., Kubik, J., ... & Reiter, H. (1997). *The social movement society: Contentious politics for a new century*. Rowman & Littlefield Publishers.

Deere, C. D., & Royce, F. S. (2009). *Rural social movements in Latin America: Organizing for sustainable livelihoods*. University Press of Florida.

Doherty, B., & Doyle, T. (2006). Beyond borders: Transnational politics, social movements and modern environmentalisms. *Environmental Politics, 15*(5), 697–712.

Glaser, B. G., & Strauss, A. L. (1967). *The discovery of grounded theory: Strategies for qualitative research* [pdf]. Zugriff unter http://www.sxf.uevora.pt/wp-content/uploads/2013/03/Glaser_1967.pdf.

Google Scholar Keyword Search Engine. (2020, May 14). https://scholar.google.com/scholar?hl=lt&as_sdt=0%2C5&q=Social+movements&btnG=.

Hess, D. J. (2016). *Undone science: Social movements, mobilized publics, and industrial transitions*. MIT Press.

Inglehart, R. (1990). Values, ideology, and cognitive mobilization in new social movements. In *Challenging the political order: New social and political movements in western democracies* (pp. 43–66). Cambridge: Polity Press.

Kivisto, P. (1984). Contemporary social movements in advanced industrial societies and sociological intervention: An appraisal of Alain Touraine's Pratique. *Acta Sociologica, 27*(4), 355–366.

Larana, E., Johnston, H., & Gusfield, J. R. (2009). *New social movements: From ideology to identity*. Temple University Press.

Lofland, J. (2017). *Protest: Studies of collective behaviour and social movements*. Routledge.

Melucci, A. (1980). The new social movements: A theoretical approach. *Information (International Social Science Council), 19*(2), 199–226.
Melucci, A. (1993). Paradoxes of post-industrial democracy: Everyday life and social movements. *Berkeley Journal of Sociology, 38,* 185–192.
Roggeband, C., & Klandermans, B. (Eds.). (2017). *Handbook of social movements across disciplines*. Springer.
Rosset, P. M., & Martínez-Torres, M. E. (2012). Rural social movements and agroecology: Context, theory, and process. *Ecology and Society, 17*(3), 17.
Rosset, P. M., Torres, M., & Elenacoaut, M. (2013, September). Rural social movements and diálogo de saberesterritories, food sovereignty, and agroecology. In *Food sovereignty: A critical dialogue international conference* (pp. 1–26). Yale University.
Sen, R., & Lee, C. H. (2015). Workers and social movements of the developing world: Time to rethink the scope of industrial relations? *International Labour Review, 154*(1), 37–45.
Sutton, P., & Vertigans, S. (2006). Islamic "new social movements"? Radical Islam, al-Qa'ida and social movement theory. *Mobilization: An International Quarterly, 11*(1), 101–115.
Tilly, C., Castaneda, E., & Wood, L. J. (2019). *Social movements, 1768-2018.* Routledge.
Touraine, A. (1971). *The May movement: Revolt and reform: May 1968–the student rebellion and workers' strikes–the birth of a social movement*. New York, NY: Random House.
Touraine, A. (1985). An introduction to the study of social movements. *Social Research, 52*, 749–787.
Touraine, A. (2002). The importance of social movements. *Social Movement Studies, 1*(1), 89–95.
Vidickiene, D. (Ed.). (2013). *Living in harmony: Inspiring stories from ecovillages*. Vilnius: Lithuanian Institute of Agrarian Economics.
Vidickiene, D., & Melnikiene, R. (2014). *Evolution of rural policy*. Vilnius: Lithuanian Institute of Agrarian Economics.
Vidickiene, D., Gedminaite-Raudone, Ž., & Simonaityte, V. (2019). *Servitization of farming: business model "product plus service" in Lithuania*. Vilnius: Lithuanian Institute of Agrarian Economics.
Waddock, S. (2017). *The difference makers: How social and institutional entrepreneurs created the corporate responsibility movement*. Routledge.
Wieviorka, M. (2005). After new social movements. *Social Movement Studies, 4*(1), 1–19.
Woods, M. (2003). Deconstructing rural protest: The emergence of a new social movement. *Journal of Rural Studies, 19*(3), 309–325.
Yin, R. K. (2003). Designing case studies. In *Qualitative Research Methods* (pp. 359–386).

CHAPTER 2

New Social Movements: Theories and Approaches

Rita Vilkė

2.1 Theoretical Background of New Social Movements

In a most general sense, social change is broadly understood as a philosophical idea, which refers to some sociocultural evolution, progress in a particular sphere, which moves the society forward compared to the previous state. Social change might refer to paradigmatic transformations in socio-economic structures (Blyth & Mark, 2002; Curran, 2020; Hassard & Cox, 2019). Among the broadly accepted examples, a transition from industrial to the post-industrial stage of development (Bell, 1976; Goncharov, 2018; Huber, 1984; Kniffin & Patterson, 2019; Loures et al., 2016) well illustrate such paradigmatic transformation of socio-economic structure. Indeed, many theories had been created and many attitudes had been developed around the world aiming to explain social changes which occurred as an outcome of effect of particular social movements. The better-known social movement theories, which were found useful at least to some extent in this research aiming to disclose the overall scientific reasoning picture of rural social movements and its impact regarding rural development paradigm shift from theoretical point of view are discussed in more detail here.

R. Vilkė et al., *Rural Economic Developments and Social Movements*, Palgrave Advances in Bioeconomy: Economics and Policies, https://doi.org/10.1007/978-3-030-71983-8_2

Relative deprivation theory. Relative deprivation theory, also known and used in a shortened version—Deprivation theory (Gurr, 1970), is one the earliest explanations of social movements, widely used in the scientific literature in the late 1960s and early 1970s, and performs evident renaissance in the 21st century again (e.g., Grasso et al., 2019; Kunst & Obaidi, 2020; Power et al., 2020, etc.). Deprivation theory finds its grassroots among the people who feel deprived of some goods or resources, and they are likely to organize a social movement to improve or defend their conditions regarding particular service, goods or comfort (Gurr, 1970; Morrison, 1978). In other words, eventhough the government institutions satisfy society's demands for timely social and political improvements, the feeling of poverty and inequality in a particular field might accelerate people to organize masses. The feeling might arise from one's comparison with other society's members or individually from oneself (one's expectations). When one's expectations exceed abilities, people organize or join social movements. Therefore motivation factors for social mobilization, in this case, are individually felt grievances and/or anger. For the context of the ongoing scientific research in this monograph concerning social movements and rural development issues, deprivation theory might be considered useful in explaining the rise of rural social movements both in times of industrial and post-industrial phases of society's development, when the particular changes in society's organization accelerated particular rural social movement to emerge. Of course, it should be necessarily stated ahead of the further analysis, that in the context of rural social movements, some sort of 'soft' form of deprivation theory might be discussed to explain the emergence of rural social movements, especially in the twenty-first century, compared to the grassroots situation of the theory application. The two huge branches of rural social movements in the twenty-first century might be observed from the 'soft' relative deprivation theory point of view. The first branch comprises industrial paradigm-driven rural social movements which emerged after the felt deprivation due to the ongoing reorganization in society's property, power, regulation, roles, and relations. Property issues accelerated the rise of social movements concerning landless agricultural workers as well as movements, protesting against the use of rural land for non-agricultural activities, such as railway, mining, bridges, hydro stations, etc. The feeling of deprivation was also a concern of small farmers, acting in the same environment as the big ones in rural areas, but seemingly remaining apart from the overall national and international government-driven agricultural support system. The industrial branch also takes into account the deprived groups which raised social movements in the fields of rural labour (trade unions), rural women and endogenous communities. In some sense deprivation was a driving force for some rural movements to emerge regarding the market regulation for exporters of agricultural production in developed countries, as well as 'fair prices for farmers' movements in developing countries, which were concerned with the redistribution of value between farmers and other agents of the supply chain. The excessive state intervention in agriculture and cut of agricultural subsidies stands for one more reason feeling deprived, which raised rural social movements concerned with the redistribution of risk between farming livelihood and other sectors. Farmer-to-farmer movement, when poor peasant farmers start teaching one another, as well as farmers' cooperative movement, emerged due to the feeling deprived regarding economic power and therefore strived to redistribute it by collective action. Finally, the overall universal attitude around the world regarding the peasant living mode, commonly considered as less qualitative than an urban one, accelerated the rise of global rural social movements (e.g., La Via Campsesina), calling to revise the peasantry/society relationships taking into account food, energy and technological sovereignty concepts and to suggest new ways to the optimal organization of society by transforming the relationships in all spheres of life (for more details, see Chapters 3 and 6). The second branch comprises post-industrial paradigm-driven rural social movements which emerged after another type of felt deprivation, which was concerned with the organization of society, everyday activity and life. Advanced skills and abilities of educated people to act in the new knowledge-based society of the twenty-first century started defining the living quality and conditions. Accordingly, separate groups of people started feeling deprived, for instance, by industrially mass-produced and provided en masse in supermarkets and everywhere less valuable food and goods. So they raised such social movements as the Slow Food movement (Hsu, 2015) and Consumers' movements, struggling to reduce the negative impact of food on health. Inline back-to-the-land movements (Wilbur, 2013), urban gardening and farming movements started emerging, aiming to diversify the activities and change the rural/urban business relationships. One more explicit surge of rural social movements emerged when urban busy people started feeling deprived by continuous everyday life tension, rush, earning for spending, and they started the movements of the new rural lifestyle, such as, ecovillages, co-housing, time bank movements, which strived to change the lifestyle and build new relations, new communities with new values, fighting the mass order of the modern society (for details, see Chapters 3 and 7). The analysed multifaceted deprivation theory application in rural discourse proves its possible use in reasoning the raise of rural social movement both in industrial and post-industrial branches. Moreover, deprivation in its broad sense uses to be a common label for rural/urban dichotomies of a social organization up till now (Dymitrow et al., 2018). Hence, it is worth considering that rural social movements of the twenty-first century in the light of deprivation theory are relatively peaceful collective actions, compared to the overall history of social movements, explained using the deprivation theory, taking into account violence and rough protests, which again start appearing in other spheres of human activity in the twenty-first century (Grasso et al., 2019; Kunst & Obaidi, 2020; Power et al., 2020, etc.). Indeed, it should be stated, that relative deprivation theory has particular limitations in overall social movement research. First, it encloses particular confusion when explaining the foundations of social movements, since the feeling of deprivation is an almost continuous feeling of all the human beings, only to a different extent: at a time people feel deprived of one good or service by different levels, and when this feeling is satisfied, they feel deprived of another one, also by different levels (Power et al., 2020). At the same time, it is quite complicated to explain why social movement occurs from one group of people, but not from another. The continuity outlines the second issue with the theory (Jenkins & Perrow, 1977) since it is circular, i.e. the only providence of deprivation is a social movement.

Mass-Society Theory. Mass-Society theory is part of the early explanations of social movements (Kornhauser, 1959). Also this theory had been recently examined by modern sociologists (Haas, 2019; Yamada, 2016). The theory argues that belonging to a particular social movement provides a sense of empowerment for individuals of large societies, who feel insignificant and socially detached. In the post-industrial society, new information and communication technologies engaged everyone into a particular mass society, and the theory became popular again in scientific research, including the examination of social movements (Nikolaeva & Fedchenko, 2016; Schroeder, 2018). Thus the renaissance of a mass society theory might be considered playing a significant role concerning the research in this monograph. The ongoing changes related to the spread of knowledge via the internet into a distanced rural areas start shaping the common consideration of rural residents as an insignificant and socially detached part of a society at large. Since education in recent decades became a basic feature of mass society (Lowe et al., 2019), the spread and application of it in rural areas start playing a significant role in rural development paradigm shift. Simply, the accessibility to the internet enabled the possibility to spread information about the ongoing variety of innovative rural initiatives worldwide, find similar-minded people, join and support particular practices with their own experiences and expertise. For instance, the ecovillages movement united similar-minded people, who previously were treated as strangers in the society they live in due to their propagated ideology towards nature cycle and their specific nature-friendly lifestyle. So, they newly created their ideology and understanding grounded communities, which became part of the ecovillages movement worldwide (for more details, please see Chapter 7). Thus those previously 'insignificant' and 'socially detached' people, sometimes from rural areas, sometimes—from crowded cities, find their same-minded passengers and become part of various social movements, which bring them back to the feeling of useful and beneficial society members, that are making a great change with their lifestyle in terms of the ongoing rural development paradigm shift.

Structural-Strain Theory. Among the early explanations of social movements the structural-strain theory (Smelser, 1963), which currently is treated next to the theories of collective behaviour (Smelser, 2011), proposes six factors for the rise of a social movement: first, structural conduciveness, when people believe that their societies have problems; second, structural strain, when people experience deprivation; third, growth and spread of a solution, when solutions for problems people are experiencing are proposed and spread; fourth, precipitating factors, when discontent requires a catalyst to turn into a social movement, normally it is a particular event; fifth, lack of social control, when the entity to be changed must be open for some change, but if the social movement is quickly repressed it might never materialize; and sixth, mobilization, which is the actual component of the movement when people do what should be done. In the context of rural development paradigm shift the given reasoning might help provide a multifaceted explanation for the rise of particular rural social movements. For instance, structural conduciveness by today's society in terms of food security and sovereignty issues might explain the birth of the agroecology movement. Agroecology has been practiced for millennia in diverse places around the world, and today we are witnessing the mobilization of transnational social movements to build, defend and strengthen agroecology as a pathway of a more just, sustainable and viable food and agriculture system (for more details, please see Chapter 7). Thus, step-by-step, structural strain theory explains the rise of the agroecology movement. Hence, there are deprivation theory attitudes, used in the structural strain theory, which is seen as shortage due to the circularity (see the relative deprivation theory above). It was also observed, that a social movement might not arise even all the Smelser's factors are satisfied, and this might confuse the research.

Resource-Mobilization Theory. Resource-Mobilization theory (Buechler, 1993; Campbell, 2018; Jenkins, 1983) argues that any motivated people should necessarily have a dispose of appropriate resources to empower their motivation into real streamed actions in a form of a social movement, i.e. resources are a core factor in this case. The theory lists different types of resources that might be used for the social movement and also stress several urgent or critical resources, which should be necessarily taken into account for social movement development and success, namely: time, money, organizational skills and social opportunities. Among the major success factors, according to the theory, are accessibility to particular resources and the ability of lead actors to utilize them, especially to modern extent and forms (Murray et al., 2020). It should be noted here, that the theory received criticism, due to the exceptional focus on resources, since there are examples around the world, including rural social movements, when all theory-listed resourced had not been mobilized, including critical ones, however, social movement examined success (i.e. U.S. Civil Rights Movement). From rural social movements' point of view, resource mobilization theory should be awarded with special attention. Plenty of initiatives in rural areas had been succeeded by mobilizing local tangible and intangible resources to make the change. It initially started with active, educated, and prospective local leaders, who, first of all, voluntarily took a responsibility to drive the demanded social change for a particular community, and developed further till the organized and institutionalized social movements. For instance, the place-based initiatives, which strived for a new understanding of the role of the peasantry, started as regional rural parliaments after the Scandinavian example in many European rural regions, gain a surge throughout Europe in the form of regional rural parliaments, which are currently united into a European Rural Parliament (for details see Chapter 6).

Political Process Theory. The Political process theory energized the examination of social movements by the very end of the twentieth century (Goodwin & Jasper, 1999). It arrived with a close content to the resource mobilization theory but emphasized political opportunities as a necessary component of social structure, which is crucially important for social movement development (Meyer, 2012). There exist the three vital components for a social movement to emerge under political process theory: first, a consciousness of insurgents, which goes back to deprivation theory, or collective sense of injustice—a mobilizing motive; second, organizational strength, which is close to resource mobilization theory, stating the strong leaders and sufficient resources; and third, political opportunities, referring to the challenge-receptive or vulnerable political system. The theory says that social movement will not succeed when there are insurgent consciousness and organizational strength, but no political opportunities. A very important advantage of the theory of political process is the model of political mediation, which refers to particular strategic choices made by social movement actors in the political context, which further leads to the discussion on the impact made by a social movement. Among the most evident application of political process theory to the analysis of rural development paradigm shift are political power-based rural social movements. For example, redistribution of property, social power for social hierarchy change, as well as added value ideas-driven rural social movements are grounded on the political process idea (for details, see Chapter 6). After uniting the joint interests in particular fields of activity, they strive to interrupt the political process and receive the gains in their demanded fields of interest. At the same time, political process theory, as well as resource mobilization theory, use to be criticized by scholars due to the ignored movement culture dimension. At the same time, political process theory, as well as resource mobilization theory, use to be criticized by scholars due to the ignored movement culture dimension.

Collective Behaviour and Collective Action Theories. The collective behaviour and collective action theories go back to the 50s of the last century, but are still exceptionally viable and broadly applied with the classical Nail Smelser's grounding as people join social movements to do a collective action because they experience strain (Smelser, 2011).

Then social movement emerge to reassure the people, that action had been done to solve the issues regarding the experienced strain (Weeber & Rodeheaver, 2003). So, collective behaviour normally refers to actions taken by a group of people together in particular circumstances. Any collective behaviour calls for the analysis of organizations, which might serve the joint purpose of members of it, or serve only the leader's interest. Hence, organizations normally fall, if the interest of its members is not fulfilled. Such organizations are expected to serve the common interest of the biggest part of members: the big farmers' common interest of direct payment legislation; the family farmers union members' interest on better small farming conditions; the stake holder's common interest of higher dividends, etc. When individuals feel, that common action would give benefits in a particular situation rather than acting individually, the presence of an organization pays off.

Value-Added Theory. The value-added theory of collective behaviour determines whether or not collective behaviour will occur. The theory argues that a specific combination of determinants facilitates and promotes collective outcomes and behaviours. The determinants of collective behaviour forms a value-added process. Value-added processes, which originated in the field of economic theory, refer to processes in which additional value is created at a particular stage of development or production. According to Knottnerus (1983), the value-added theory asserts that determinants to collective behaviour combine according to a predictable pattern. Collective behaviour requires the appearance of the determinants in a logical and predictable order; specifically, the theory asserts that six social conditions or 'determinants are required for the development of a social movement: structural conduciveness, structural strain, generalized beliefs, precipitating factors, mobilization of participants, and social control'. As it is initially seen from the already made overview of social movement theories, the value-added theory has particular similarities to the already discussed structural-strain theory. The main distinctive feature is, that value-added theory proposes, that values, followed by norms, roles, and facilities, composes the most important factors that influence social behavior and collective action, and the values are the foundation for social system integration and institutionalized action. Thus the value-added theory helps explain, how grievances become generalized beliefs and compose the grassroots for a social movement to arise.

Frame Analysis Theory. The frame analysis theory was born after the propositions made by Erving Goffman (1974). This theory emphasize the understanding of the way in which social movements and actors of social movement create and use meaning, or how events and ideas are framed. Thus, this meaning work has become a keyway in which social movements are understood and analysed. As stated by Benford and Snow (2000): 'framing processes have come to be regarded, alongside resource mobilization and political opportunity processes, as a central dynamic in understanding the character and course of social movements' (p. 612). It is important to consider that people frame experiences in order to organize and understand the world around them. The theory emphasize, that previous social position and experiences help people interpret the world. Every social interaction that occurs is understood through a frame of reference within which people react based on their perception of the situation and the way they perceive the people with whom they are interacting (Gofman, 1974). Hence, collective action frames in the analysis of social movements are to trigger people to action when they are brought together. According to Benford and Snow (2000), 'Collective action frames are action-oriented sets of beliefs and meanings that inspire and legitimate the activities and campaigns of a social movement organization (SMO)' (p. 614). Hence, there is a distinction made in the theory regarding the overall social movement, and a particular social movement organization (SMO), which is considered as a smaller part of the social movement, which is often sponsored with various resources from a social movement. SMOs normally set up collective action frames to create a set of meanings which will inspire people to act collectively towards some goal. According to Benford and Snow (2000) social movement framing analysis focuses on four broad areas, further including subcategories of analysis: (a) the creation and use of collective action frames, (b) framing processes, (c) opportunities and constraints and (d) the effect of framing on movement outcomes and other processes (p. 612–613). There had been sole attempts in scientific studies to utilize frame analysis theory for rural social movements (e.g. Mooney, 2000; McKeon, 2013), however, it had not beed awarded greater exploitation for research in the field. Frame analysis theory holds multiple options, namely: collective action frames (Gamson & Meyer, 1996), diagnostic framing (Gamson, 1995; Jenness, 1995), prognostic and motivational framing (Benford & Snow, 2000). This might be taken as a preposition for further research concerning rural social movements, still unused in this monograph due to the different purposes and selected methodology of this research.

Culture Theory. Culture theory fulfills both previously outlined political process and resource-mobilization theories (Johnston & Klandermans, 1995; Hart, 1996; Armstrong & Bernstein, 2008), by extending them in two ways. First, it emphasizes the importance of movement culture, which is very relevant in terms of post-industrial change-driven social movements (see Chapter 7) and overall rural development paradigm shift. For instance, particular new rural social movements raised specifically intending to create different culture-based communities and or/settlements, which completely differ from the existing ones (e.g. ecovillages). Second, it attempts to address the free-rider problem. Both resource-mobilization theory and political process theory include a sense of injustice in their approaches, whereas culture theory brings this sense of injustice to the forefront of movement creation by arguing that, for social movements to successfully mobilize individuals, they must develop an injustice frame, which indirectly comes after Benford and Snow (2000) framing theory. An injustice frame is understood as a collection of ideas and symbols that illustrate both how significant the problem is as well as what the movement can do to alleviate it. For instance, agroecology rural social movement raised due to the observed unjust treatment with the environment when satisfying the demand for human food and fiber. Moreover, in emphasizing the injustice frame, culture theory also addresses the free-rider problem. The free-rider problem refers to the idea that people will not be motivated to participate in a social movement that will use up their personal resources (e.g. time, money, etc.) if they can still receive the benefits without participating. A significant problem for social movement theory has been to explain why people join movements if they believe the movement can/will succeed without their contribution. Culture theory argues that, in conjunction with social networks being an important contact tool, the injustice frame will motivate people to contribute to the movement. Such a situation is illustrated in this monograph with particular cases from the analyzed change-focused post-industrial rural social movements (see Chapter 7).

New Social Movement Theory. New social movement theory argues that contemporary social movements, which are often characterized as holding the 'exlosive power' in quickly mobilizing the masses and resources (Roggeband & Klandermans, 2017; Waddock, 2017, etc.) are performing collective action in markedly different ways than those traditional social movements (Crozat et al., 1997; Larana, 2009; Castells, 2015; Lofland, 2017; Tilly & Wood, 2015).

Commonly sociologists use new social movement theory to explain the role of social movements in post-industrial societies. New social movement theory refers to a new paradigm of social movement activity and above discussed collective action. New social movements (further - NSMs) are characterized by strategies, goals and membership, which are distinct from traditional social movements (Crozat et al., 1997; Larana, 2009). Theorists and scholars (e.g., Roggeband & Klandermans, 2017) NSMs as arising from numerous channels in society. For example, NSMs are seen as expressions of civil society's desire for structural change and arise from the growing importance and ubiquity of information in our increasingly knowledge-based society. NSMs are also seen as an inevitable outcome of changing social, economic and political relationships in the post-industrial society. NSMs are therefore considered as movements for change, based on the desire for structural reform rather than revolution, and they do not attempt to dismantle the existing political and economic systems. NSMs helps to explain the changing forms of political organization and the shifting relations between public and private spheres in post-industrial societies (Lentin, 1999). New social movement theory dominates current social movement research and allows for the study of macro external elements and micro internal elements (Fuchs, 2006).

New social movement theory argues that NSMs are distinct from other traditional social movements. Traditional social movements, such as labour movements, tend to be engaged in class conflict while NSMs (e.g., anti-war, environmental, civil rights, feminist movements) are more engaged in political and social conflict. Moreover, traditional social movements tend to focus on economic concerns and inequalities, whereas members of new social movements are most often from a segment of society referred to as the new middle class. NSMs encourage members to engage in lifestyle changes, tend to have supporters rather than members and are characterized as loosely organized networks—it is very common to the ongoing research in this monograph—rural social movements. So, NSMs differ from protest groups or movements as they often desire to see change on a global scale as opposed to the single issues taken on by protest groups.

In suming up, it should be highlighted, that scientific discussions stress how greatly the understanding of a social change varied throughout history (Haferkamp & Smelser, 1992). By the end of the twentieth century, Herman Strasser and Suzan Randall done the review of theories of social change and outlined particular key attributes of change: 'magnitude of change, time span, direction, rate of change, amount of violence involved' (1981, p. 16). More than a decade later Hans Haferkamp and Neil Smelser (1992, p. 2) proposed a threefold each other fulfilling simplified meta-framework for models of social change: first, social change structural determinants (population changes, dislocation changes due to the war, strains and contradictions); second, social change processes and mechanisms (precipitating mechanisms, social movements, political conflict and accommodation, and entrepreneurial activity); and third, social change directions (structural changes, effects and consequences). In other words, this simplified meta-framework of social change discloses the relationship among the structural determinants of social change, which are 'the accumulated consequences of previous sequences of change' (Haferkamp & Smelser, 1992, p. 3).

Integration of the most recent scientific findings concerned with social change with new social movements theory help distinguish the several sources for social change, which are relevant in the analysis of new social movements. One group of sources for social change is concerned with systemic factors that boost the social change due to the maturity of a particular state of the art, for instance, flexible and stable government, sufficient resources, a diverse social organization of society, etc. Therefore social change occurs when the existing state is over-fulfilled with appropriate factors, and there is a need for progress or evolution. Another group of sources, which is very characteristic to recent transformations around the world, outlines the specific unique and random factors for social change, such as the presence of the specific groups of people, believes, lifestyle, area, climate, weather, etc. In this case, social change occurs when there is a need to react to a specific unique situation or factor, which does not satisfy particular individuals, groups of actors, community or, moreover, society.

The outlined distinction among the factors for social change, compose the grassroots for theoretical background helpful in the analysis of new social movement phenomena. At the beginning of the twenty-first century, the belief, that social change is caused or determined by a single factor, which stands for reductionist and determinist theories, is often criticized due to its outdated incapability to go in line with the modern sustainability thought, which undoubtedly has a multifaceted nature (Gasparatos et al., 2008). Recent scientific discussions most often agree, that social change in current times occurs in the interaction of both groups of factors—systemic and particular unique factors (Heaphy, 2007; McMichael, 2016; Shackman et al., 2002; Vogt, 2016), which represent the interactionist branch of theories. The interaction of multiple groups of factors in every unit of analysis, e.g. every case of social change, and especially in a form of new social movements, is unique and complex, and thus calls for the particular combination of research attributes to be activated in such area of research, putting the qualitative measures in the first place.

2.2 Attributes of New Social Movements

The necessary part of many scientific investigations, concerning the NSMs, are their attributes. The review of multiple scientific literature sources suggest the folowing most relevant attributes of NSMs: the ideology and goals, tactics, organizational structure, and participants.

Ideology and goals. The ideological outlook use to be considered a central distinct characteristic of NSMs, highlighted in multiple scientific studies (e.g., Dalton et al., 1990; Pichardo, 1997; Oliver & Johnston, 2000; Ryan, 2013; Allen et al., 2017; etc.). It is often stated in research, that ideological outlook is grassroots for further developments regarding the overall social movement theory and its transformations. For instance, the NSM paradigm states that currently social movements represent a fundamental break, compared to social movements from the industrial stage of development. Recently, NSMs are more often focused on quality of life, various lifestyle concerns, rather than economic redistribution of property, power, hierarchy, etc. Thus, NSMs question the wealth-oriented materialistic goals of industrial societies. They also call into question the structures of representative democracies that limit citizen input and participation in governance, instead advocating direct democracy, self-help groups and cooperative styles of social organization (Pichardo, 1997). Researchers observe, that the values of NSMs mostly centre on autonomy and identity (e.g., Offe et al., 1985). Hence, identity claims are the most distinctive feature of NSMs (Kauffman, 1990), although all previous movements can also be described as expressing identity claims (Pichardo, 1997). The focus on identity is considered unique in contemporary movements because, as stated by Kauffman (1990), 'identity politics also express the belief that identity itself-its elaboration, expression, or affirmation-is and should be a fundamental focus of political work. In this way, the politics of identity have led to an unprecedented politicization of previously non-political terrains' (p. 67).

However, as it might be observed from literature review concerning the impact of identity on participation in a social movement, this context received minor attention in scientific research. Among one of the sound empirical inputs concerning the issue was done by Klandermans (1994). He examined the Dutch peace movement, focusing on collective identities and their variety by different organizational memberships. It was predicted defection of participants from the movement after occurred identity conflicts. One more unpublished work, concerning the issue, have examined the role of personal identity in relation to participation in the environmental movement (Pichardo, 1997). There was found a significant, but not strong, association between self-reported environmental identity and participation both in conventional social movement activities (event participation, organizational membership, movement contributions) and in everyday behaviours (conserving energy and water, using alternative transportation and purchasing products made from recycled materials). Researchers state, that more empirical work on the connection between identity, at all its levels, and movement participation needs to be done (Pichardo, 1997).

Another specific ideological feature of NSMs, observed in scientific research, is its self-reflexive character, emphasizing that participants are constantly questioning the meaning of what is being done (Cohen, 1985; Gusfield, 1994; Melucci, 1994). Often the given example of this is the consciousness-raising groups, which are characteristic to feminist movement (Pichardo, 1997). It is important to state, that this specific ideological feature help define types of NSMs. Observations suggest, that the unique ideological orientation and self-reflective character leads to the choices of tactics, structures, and participants in NSMs. Hence, the in-depth use of this feature in the analysis of rural paradigm shift and rural social movements is further developed in Chapter 3 of this book.

Tactics. Plenty of scientific studies in the 21st century suggest, that collective identities of social movements are commonly reflected in NSMs tactical choices (Tarrow, 1994; Pichardo, 1997; Rojas, 2006; Smithey, 2009; Feinberg et al., 2017; Doherty & Hayes, 2019). At the end of the 20th century, it was believed that NSMs preferred to remain outside of normal political channels, employing disruptive tactics and mobilizing public opinion to gain political leverage (Pichardo,

1997). The use of highly dramatic and preplanned forms of demonstrations use to be repleted with costumes and symbolic representations (Tarrow, 1994). Recent scientific findings outlines, that the exploitation of extreme tactics leads to the social movement members' willingness to espoused extreme positions (Feinberg et al., 2017). However, oppositely to the belief, that extreme tactics would serve to strengthen the ideology, help mobilize resources and reach institutionalization, in line with other aims, social movements more often 'expose' and diminish just after the adoption of such risky tactics (Doherty & Hayes, 2019).

It is observed, that NSMs do involve in politics, and often strive to become institutionalized (Pichardo, 1997). Some NSMs have become integrated into the party system and gained regular access to regulatory, implementation and decision-making bodies, while others have formed political parties that regularly contest for electoral representation (McAdam & Tarrow, 2018). The scientific findings give evidence that a number of Green parties are prominent in Europe, with several having local manifestations in the United States (Müller-Rommel, 1985). However, no direct correspondence had been observed between supporters of NSMs and those who vote for Green parties (Müller-Rommel, 1985). Pichardo (1997) outlines, that the NSM paradigm recognizes the fact of non-existing truly distinctive tactical style of NSMs; public opinion and anti-institutional politics have been more prominent additions to the repertoire of social movements. Hence, recent scientific findings outlines, that the use of extreme tactics lead to the social movement members' willingness of espoused extreme positions (Feinberg et al., 2017). However, oppositely to the belief, that extreme tactics would serve to strenthen the ideology and mobilize resources, or serve other movement's aims, social movements more often decline and diminish just after the adoption of such tactics (Doherty & Hayes, 2019). Further chapters of this book give an outline regarding the state of rural social movements tactics and institutionalization, as well as political involvement, for instance, in such a consolidated form as European Rural Parliament (for details, see Chapter 6).

Structure. In scientific research, the importance of NSMs' organizational structure among the other crucial attributes has been addressed more than fifty years ago (Gerlach & Hine, 1968). By the end of the 20th century, based on Pichardo (1997) observations, NSMs organized themselves in a fluid non-rigid style avoiding the dangers of oligarchization. At the same time they tended to rotate leadership, voted communally on all issues and had impermanent ad hoc organizations (Offe, 1985). Structure-specific features of NSMs had been further characterized by taking into account the issues of speaking out against bureaucratic attitudes, arguing against what they perceive as the dehumanizing nature of modern bureaucracy. Due to these specific features, NSMs demanded and, therefore, create structures that are more responsive to the needs of individuals—nonhierarchical, decentralized and open. Thus the willingness to avoid becoming co-opted or deradicalized, as suggested by the lessons learned from a historical pathway, played a motivational role for greater flexibility in structures. It should be stated here, that the ideal organizational structure, which meets all the outlined features, is far from a common attribute of NSMs. For instance, various environmental movements, the National Organization of Woman tend to be more traditionally centralized, hierarchical forms of organization (Shaiko, 1993). It is observed by scholars, that in a timeline, especially after the knowledge and information appeared as crucial factors in the post-industrial societys' development, NSMs tend to flatter their organizational structures and start using more project or network-based forms, including modern social networks and other media forms (Ray & Tarafdar, 2017; Schneiberg & Lounsbury, 2017; Tremblay et al., 2017; Van Dyke & Amos, 2017; Leong et al., 2020; etc.). For example, Ray and Tarafdar (2017) issuing the impact of social media on social movements as highly debated and not clearly understood, did a research study and looked at the role of Twitter in a social movement that emerged in India, post a violent gang-rape observed, taking into account the structure of the movement as one of the dynamic features. It was found, that Twitter impacted important structural components of the social movement, including the organizational structure. Thus, in the ongoing research NSMs' structure should be necessarily taken into account from the knowledge and information age point of view, as it is further done in this monograph.

Participants. Scientific literature, close to the end of the 20th century, suggested mainly the two basic views regarding the participants of NSMs: who they are and why they join (Pichardo, 1997; Cotgrove & Duff, 1981; Lowe & Goyder, 1983; Rudig, 1988). The first is a postindustrial economy-based view, by placing the base of support for NSMs within the 'new' middle class, which emerged as social stratum employed in the nonproductive sectors of the economy (Cotgrove & Duff, 1981; Lowe & Goyder, 1983; Rudig, 1988). Research on the rise of the new middle class within postindustrial society seems to establish the credibility of this social phenomenon (Brint, 1994). But, as observed by Pichardo (1997), NSM theorists go a step beyond, by arguing that this stratum produces the chief participants of NSMs because they are not bound to the corporate profit motive nor dependent on the corporate world for their sustenance. Normally such participants are highly educated and work in the areas, that are highly dependent upon state expenditures, for instance, academia, the arts and human service agencies. The education factor is very issue-specific to the rural social movements of the 21st century, such as the ecovillages movement, which are established by society's members, who are experienced and/or educated, striving to create a 'new world'—a place for a community with its own culture and lifestyle and live in harmony with nature, at the same time serving all their internal needs, which are necessary for human beings (for details, see Chapter 7). The second view on the participants of NSMs is related to the conflict over the control of work (Kriesi, 1995; Schneiberg & Lounsbury, 2017). From this point of view, participants of NSMs are defending their professionalism, based on expertise and skills, against the attempts on their work autonomy, done by colleagues, involved in administration at a large. Concerning the context of rural social movements, a good example is the movement of a new role of the peasantry, which is guided by the idea, that the primary function of agriculture and rurality had faded decades ago. Currently, the peasantry is a synonym of harmonious collaboration between human beings and the nature as highest healthy and happy living standard. Soil digging, livestock growing, gardening, and other multiple activities, which use to be considered as core rural activities, guiding the peasantry, already diminished (for more details, see Chapter 6).

Throughout the literature review regarding the issue of participants of NSMs, it was observed, that such a twofold categorization is too limited regarding the issue (e.g., Pichardo, 1997; Schneiberg & Lounsbury, 2017; McAdam & Tarrow, 2018, etc.). The other this research-relevant view of the participants of NSMs is that they join and unite because of a particular social concern, instead of the class boundaries. It is, at the same time, an ideological, rather than ethnic, religious or class-based community (Pichardo, 1997). For example, West German Greens are defined by common values rather than a common structural location and corresponds to the 'catch-all' party (Arato & Cohen, 1984; Halfmann, 2018).

There also exist slightly different view regarding the NSM participants (Offe, 1985). He argues that they are drawn from three sectors: the new middle class, elements of the old middle class (farmers, shop owners and artisan-producers), and a 'peripheral' population consisting of persons not heavily engaged in the labour market (students, housewives and retired persons). A number of studies of the peace movement in various countries have demonstrated an equally diverse set of participants (Kaltefleiter & Pfaltzgraff, 1985). Diani and Lodi (1988) show that within the Milan ecology movement, several different currents attract somewhat different sets of participants.

However, scientific evidence regarding the variety of studies NSMs most often ignore the rules of any classification, proposed in the research. For instance, studies of environmental movements reveal that NSM participants are drawn primarily from two populations: The 'new' middle class is one; the other is geographically bound communities that are being directly affected by the negative externalities of industrial growth. Participants in such classification are the more ideologically committed middle class as well as communities that protest the siting of hazardous waste sites, landfills and waste incinerators, or chemical and/or radiation poisoning of the local environment (Walsh, 1981; Pichardo, 1997). The old middle class typically is also involved in regional issues (Touraine, 1981). In short, the participants of environmental movements do not draw significantly from outside the white middle class unless there is some motivating, geographically based, grievance. For example, minority communities have rarely participated in the environmental movement, except in protest over the placement of unwanted waste facilities (Falcone et al., 2020). This kind of placement has been referred to as environmental racism, a term grounded in the belief that decisions about locating hazardous waste sites ignore the interests of minority groups.

The outlined attributes of social movements suggest, that there are plenty of possibilities and options to analyse social movements, starting from the ideology and goals, then considering the tactical issues, further—organizational structure, and finally—participants and the reasoning of participation. At the same time, there is a possibility to focus on a selected attribute and implement comparative analysis of several social movements, especially when particular topicality is a uniting factor. Hence, the outlined attributes do not refer to any lifecycle of NSMs, which is particularly one of the most exceptional characteristics of a movement, compared to other more durable collective behaviour types.

Therefore, another,—lifecycle perspective, regarding the social movements might help explain the rural development paradigm shift and the role of rural social movements in the phenomena under research.

2.3 Development Stages of Social Movements

NSMs received great attention from the scientific community and were well documented from different perspectives (e.g. Buechler, 1995; Inglehart, 1990; Melucci, 1980; Touraine, 1971; Wieviorka, 2005, etc.). The distinction of stages of NSM development had been firstly proposed by famous sociologist Herbert Blumer and might be considered useful in the context of this research—when exploring and explaining the prospects of innovative rural development ideas that change the state of the ongoing rural development thus corresponding to a rural development paradigm shift (see Fig. 2.1).

From various sources of literature, the overall process of NSM consists of several common stages (Castells, 2015; Crozat et al. 1997; Inglehart, 1990; Larana, 2009; Lofland, 2017; Tilly & Wood, 2015; Imhonopi

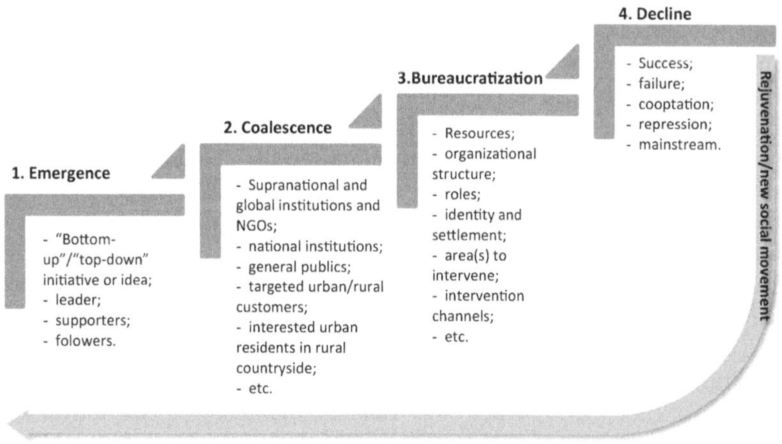

Fig. 2.1 Lifecycle stages of social movement development (*Source* Elaborated by author, using Inglehart [1990], Buechler [1993], Crozat et al. [1997], and Tilly and Wood [2015])

et al., 2013), which, if disclosed, help to form the analytical framework for the research of this monograph (see Chapter 1) .

Stage 1: Emergence. Any NSM starts with an idea, which might be recognized as bottom-up initiative of a particular leader or community. Various sources of literature suggest the same title for the first stage of a NSMs, i.e., 'Emergence'. Blumer proposed the description of a 'social ferment' for this initial stage of NSMs (De la Porta & Diani, 2006). Considering its essence at this stage, social movements might be decribed as widespread discontent rather that any preliminary organization (Macionis, 2001). The initially featured characteristics, that might already signalize the emergence of a new social movement, might be recognized after going back to the above-outlined theories (see section 2.1.). The very initial potential for the emergence of a new social movement might be recognized from the individual, rather than a collective raise of particular actions. Normally it happens by spectating a small group of potential future movement's participants: the leader, or a group of people, unhappy with the existing policy or social issues, their quality of life or social status conditions, their close or remote environment, etc. At this initial stage the increase in person's (small group's)—future leader's of a new social movement, activism is normally observed, there are no evident signs of collective action. A person may comment to friends and family that he or she is dissatisfied with conditions or may write a letter to the local newspaper or representative. Hence, at this stage the actions are not strategic, nor collective. Further, there may be an increase in media coverage of negative conditions, unpopular policies or other social concerns, which contributes to the general sense of discontent. This early stage can also be considered within a specific social movement organization (SMO), which had been already addressed (see in Section 2.2.). An example of the context-specific SMO in this book might be the Lithuanian Rural Parliament, which is further represented in European Rural Parliament (for details—see Chapter 6). Among the worldwide examples, the Student Non-violent Coordinating Committee (SNCC) well corresponds to the SMO, which was one of the many social movement organizations that was organized during the American Civil Rights Movement (Imhonopi et al., 2013). SMO and its members at the emergence stage serve as agitators, who raise consciousness around issues and help to develop the sense of discontent among the general population. An example of this stage would be the early 1950s for the Civil Rights Movement.

Stage 2: Coalescence. When the idea had emerged, it is necessary to find support for it. Thus coalesce stage begins. Normally coalesce stage attracts different groups of stakeholders, using a variety of tools and techniques. In previous times various media channels used to be recognized as the quickest way to mobilize necessary resources for NSM (Roggeband & Klandermans, 2017). In current times, networked society holds explosive social network power which serves as a quickest way to spread any news in a second throughout the world (Hanspeter et al., 2015).

This stage of NSMs cycle is specific due ot the issue of mobilization. There are observations in scientific research when the widespread mobilization does not happen despite the existing social unrest and discontent (Macionis, 2001). For instance, people in a community may complain to each other about a general injustice, but they do not come together to act on those complaints and the social movement does not progress to the next level. The second stage is also known as the 'popular stage', which is characterized by a more clearly defined sense of discontent. It is no longer just a general sense of unease, but now a sense of what the unease is about and who or what is responsible for that. Also, at this stage mass demonstrations may occur in order to display the social movement's power and to make clear demands. Most importantly this is the stage at which the movement becomes more than just random upset individuals. Most important feature here is a strategic outlook and organization. It is worth again mentioning the American Civil Rights Movement as an example, when the initial emergence was further lead with a series of high-profile campaigns, which sought to highlight the plight of African Americans in the segregated South (Imhonopi et al., 2013). After many years of successful, but hard-fought campaigns and strong leadership, the movement became a more prominent political force.

Stage 3: Bureaucratization. Bureaucratization phase is mainly concerned with resources and its allocation, organizational structure and roles, identity and settlements, areas of intervention and intervention channels, and other relevant bureaucratization issues and procedures. It is a very important stage of NSMs development since the originality and excellence achieved throughout the process of emergence and coalesce, mature in this bureaucratization phase, and determines further development of NSM.

The collected evidence of the scientific literature concerning this stage of NSMs stresses the characteristics of higher level of organization, and developed strategies. In case the social movement had already succeded by raising awareness in the previous stages, the time comes when coordinated startegy becomes a necessity. In case NSM had been previously spread into several SMOs, the coordinated strategy need to be set across all SMOs (Imhonopi et al., 2013). Tho carry out the movement goals, it starts being necessary to keep staff with specialized knowledge (e.g., administrative, management, accounting, etc.) for everyday operations, coordination and asistance. There is no longer possible for social movements at this stage to rely upon masses and/or inspirational leaders and keep progressing since only the trained staff might accommodate the already mobilized all kinds of resources. It is stated in literature, that in this phase their political power is greater than in the previous stages in that they may have more regular access to political elites. The multiple observations are made how many or newly emerged social movements fail to bureaucratize in the described way, since it is very difficult for members to sustain the emotional excitement, which is necessary. At the same time the continued mobilization becomes too demanding for participants. Finally, the formalization often shows that inspiring and enthusiastic volunteers are not ready enough to devote more of their knowledge, time, and other resources to run the movement further, and then the paid staff fully satisfy the movement's needs in fulfilling the goals (Macionis, 2001). In general, the most evident example of the successful bureaucratization phase is the recently exceptionally active gay rights movement, which performs success worldwide—they timely moved from agitation and demonstrations to having many formal organizations that now work towards the goals of the gay rights movement. Some of these organizations include the Human Rights Campaign and the Gay and Lesbian Alliance Against Discrimination (GLAAD). Another good example from the rural social movements is the peasants' rights movement 'La Via Campesina' (for details, see Chapter 6), which, if not timely bureaucratized all new role of peasantry-concerned SMOs, would have most likely faded away and their demands would have gone unmet.

Stage 4: Decline. Various sources of literature (e.g. Castells, 2015; Lofland, 2017; Tilly & Wood, 2015, etc.) stress that any NSM comes to the decline phase regardless of the output from bureaucratization phase, and only the duration of NSM in time vary. So, further scenarios of NSMs life cycle might be diverse. NSM might be rewarded by success or failure, co-optation or repression, of becoming a mainstream, but only for a particular period of time.

It is important to consider here, that the decline, does not necessarily mean failure for social movements though. Instead, Miller (1999) argues, there are four ways in which social movements can decline, hence other (e.g. Macionis, 2001) suggest five ways, namely:

- Repression—one of the most vulnerable and aggresive NSMs decline outcomes, considered to occur when authorities, or agents acting on behalf of the authorities, use measures, including the violent ones, to control or destroy a social movement.
- Cooptation—occurs when movement leaders come to associate with authorities or movement targets more than with the social movement constituents.
- Success—some movements decline because they are successful, i.e. smaller, localized movements with very specific goals often have a better chance at outright success.
- Failure—has multiple reasons, but among the most common is the one when the organization is not able to handle the rapid expansion that occurred because of their success; due to organizational strain, it collapsed into different factions.
- Mainstream—when the goals or ideologies are adopted by the mainstream, and there is no longer any need for a movement.

The decline phase is quite often considered as a kind of failure of NSMs, but it is a misleading proposition for many above outlined outcome scenarios. Moreover, going back to the theories of social movements, it is worth remembering, that the final aim of this kind of collective action—social movement,—is to make social change, and when the change is made, based on the lifecycle approach, it is considered, that the movement had reached its decline phase.

So the overviewed life cycle approach towards NSMs suggests that the movement finishes with one of the outcomes from the decline phase, since the desired impact had been already reached and made through the implemented actions, defined by the movements' strategy. Hence, the decline further might lead to rejuvenation or birth of a new social movement. Taking into account the recent observations (e.g. Jessop et al., 2013; Vasin et al., 2017; Bennett & McWhorter, 2019; etc.), social movements start being analysed as learning organizations, that are generating social innovations. Hence it is a separate, or evolutionary, furthermore—advanced approach towards the development of NSMs, which might shape the consideration of the rural development paradigm shift in future research. Of course, there might rise a natural question, whether the social movement might be considered as such, in case it does not go through the whole lifecycle? In this research by taking a lifecycle approach, there is made a presumption, that every movement goes through a lifecycle, only the expression and/or fulfillment of each stage might differ. And this is further taken as a guiding principle for case studies of the monograph (Chapter 6 and Chapter 7).

Summing up the outlined theoretical background for NSMs it might be considered, that plenty of theories and approaches had been developed to explore and explain the grassroots, the attributes and the lifecycle of social movements, starting from the second half of the last century till nowadays. There had been great attempts of scientists to make a distinction between the old, sometimes called classical social movement from the new ones—NSMs. The outlined theories and approaches with given examples of its application in broader and in the rural discourse elucidate, that they might successfully serve the analysis of both—old and new social movements. And the distinctive feature mainly rests in the focus of a particular social movement. So, by making a distinction between the industrial and post-industrial stage of society's development, we consider the change in focus of organization and actions. Thus, in an evolutionary timeline, it might be observed, that the goals and focus of a social movement is a more reasonable distinctive feature, that corresponds to grouping of old and new social movements. In this book, the initial presumption is made, that rural social movements, that adds here to the explanation of the rural development paradigm shift, advocates the two key streams: the industrial rural paradigm grounded rural social movements, and those industrial paradigm shift-focused rural social movements. Both types correspond to the reconstruction of life quality and lifestyle in rural areas, so are considered being NSMs. Further Chapter 3 puts the grassroots for a rural development paradigm shift by more in-depth analysis and discussion regarding the outlined issues.

References

Allen, S., McCright, A. M., & Dietz, T. (2017). A social movement identity instrument for integrating survey methods into social movements research. *SAGE Open, 7*(2), 2158244017708819.

Arato, A., & Cohen, J. (1984). Social movements, civil society, and the problem of sovereignty. *Praxis International Oxford, 4*(3), 266–283.

Armstrong, E. A., & Bernstein, M. (2008). Culture, power, and institutions: A multi-institutional politics approach to social movements. *Sociological Theory, 26*(1), 74–99.

Bell, D. (1976, May). The coming of the post-industrial society. *The Educational Forum,* 40(4), 574–579. Taylor & Francis Group.

Benford, R. D., & Snow, D. A. (2000). Framing processes and social movements: An overview and assessment. *Annual Review of Sociology, 26*(1), 611–639.

Bennett, E. E., & McWhorter, R. R. (2019). Social movement learning and social innovation: Empathy, agency, and the design of solutions to unmet social needs. *Advances in Developing Human Resources, 21*(2), 224–249.

Blyth, M., & Mark, B. (2002). *Great transformations: Economic ideas and institutional change in the twentieth century*. Cambridge University Press.

Boren, M. E. (2001). *Student resistance: A history of the unruly subject*. New York: Routledge.

Brint, S. (1994). Sociological analysis of political culture: An introduction and assessment. *Weil, FD and Gautier M. Political Culture and Political Structure: Theoretical and Empirical Studies, 2*, 3–41.

Buechler, S. M. (1993). Beyond resource mobilization? Emerging trends in social movement theory. *The Sociological Quarterly, 34*(2), 217–235.

Buechler, S. M. (1995). New social movement theories. *Sociological Quarterly, 36*(3), 441–464.

Campbell, A. (2018). *A Comparison of unschoolers and evangelical homeschoolers in Central Texas through the lens of resource mobilization theory*.

Castells, M. (2015). *Networks of outrage and hope: Social movements in the Internet age*. Wiley.

Cerroni, A. (2018). Steps towards a theory of the knowledge-society. *Social Science Information, 57*(2), 322–343.

Churchill, W., & Wall, J. V. (1990). *The COINTELPRO papers: Documents from the FBI's secret wars against domestic dissent*. Boston: South End Press.

Cohen, S. (1985). *Visions of social control: Crime, punishment and classification* (pp. 145–148). Cambridge: Polity Press.

Cotgrove, S., & Duff, A. (1981). Environmentalism, values, and social change. *British Journal of Sociology*, 92–110.

Crozat, M., Hipsher, P. L., Katzenstein, M. F., Keck, M. E., Klandermans, B., Kubik, J., ... & Reiter, H. (1997). *The social movement society: Contentious politics for a new century*. Rowman & Littlefield Publishers.

Curran, S. R. (2020). Green-or rose-coloured lenses for Globalization Matters? Transdisciplinary epistemic practices and paradigmatic transformations in ecologies and equalities. *Globalizations, 18*, 1–12.
Dalton, R. J., Kuechler, M., & Bürklin, W. (1990). The challenge of new movements. *Challenging the political order: New social and political movements in Western democracies*, 3–20.
De la Porta, D., & Diani, M. (2006). *Social movements: An introduction* (2nd ed.). Malden, MA: Blackwell Publishing.
Diani, M., & Lodi, G. (1988). Three in one: Currents in the milan ecology movement. *International Social Movement Research, 1*, 103–124.
Doherty, B., & Hayes, G. (2019). *Tactics and strategic action*. Oxford: Wiley Blackwell.
Dymitrow, M., Biegańska, J., & Grzelak-Kostulska, E. (2018). Deprivation and the Rural-Urban Trap. *Tijdschrift voor economische en sociale geografie, 109*(1), 87–108.
Falcone, P. M., D'Alisa, G., Germani, A. R., & Morone, P. (2020). When all seemed lost. A social network analysis of the waste-related environmental movement in Campania, Italy. *Political Geography, 77*, 102114.
Feinberg, M., Willer, R., & Kovacheff, C. (2017). *Extreme protest tactics reduce popular support for social movements* (Rotman School of Management Working Paper 2911177).
Freeman, J., & Johnson, V. (1999). *Waves of protest: Social movements since the sixties*. Lanham, MD: Rowman & Littlefield.
Fuchs, C. (2006). The self-organization of social movements. *Systemic Practice and Action Research, 19*(1), 101–137.
Gamson, W. A. (1995). Constructing social protest. *Social Movements and Culture, 4*, 85–106.
Gamson, W. A., & Meyer, D. S. (1996). *Framing political opportunity*.
Gasparatos, A., El-Haram, M., & Horner, M. (2008). A critical review of reductionist approaches for assessing the progress towards sustainability. *Environmental Impact Assessment Review, 28*(4–5), 286–311.
Gerlach, L. P., & Hine, V. H. (1968). Five factors crucial to the growth and spread of a modern religious movement. *Journal for the Scientific Study of Religion*, 23–40.
Goffman, E. (1974). *Frame analysis: An essay on the organization of experience*. Harvard University Press.
Goncharov, V. N. (2018). Traditional, industrial and post-industrial societies: social and philosophical analysis in the context of historical development. In *Society, culture, personality in modern world* (pp. 14–17).
Goodwin, J., & Jasper, J. M. (1999, March). Caught in a winding, snarling vine: The structural bias of political process theory. *Sociological Forum, 14*(1), 27–54. Kluwer Academic Publishers-Plenum Publishers.

Grasso, M. T., Yoxon, B., Karampampas, S., & Temple, L. (2019). Relative deprivation and inequalities in social and political activism. *Acta Politica, 54*(3), 398–429.

Gurr, T. R. (1970). Sources of rebellion in Western societies: Some quantitative evidence. *The Annals of the American Academy of Political and Social Science, 391*(1), 128–144.

Gusfield, J. (1994). Political processes and institutions-making It work: The prostitutes' rights movement in perspective by Valerie Jenness. *Contemporary Sociology, 23*(3), 379.

Haas, M. (2019). Mass society paradigm. In *Why democracies flounder and fail* (pp. 31–65). Cham: Palgrave Macmillan.

Haferkamp, H., & Smelser, N. J. (Eds.). (1992). *Social change and modernity* (pp. 1–33). Berkeley: University of California Press.

Halfmann, J. (2018). 3. Social change and political mobilization in west Germany. In *Industry and Politics in West Germany* (pp. 51–86). Cornell University Press.

Hanspeter, K., Koopmans, R., Duyvendak, J. W., & Giugni, M. G. (2015). *New social movements in western Europe: A comparative analysis*. Routledge.

Hart, S. (1996). The cultural dimension of social movements: A theoretical reassessment and literature review. *Sociology of Religion, 57*(1), 87–100.

Hassard, J., & Cox, J. W. (2019). Philosophy of science, social theory, and organizational analysis: Paradigmatic transformations since the postmodern turn. In *The Emerald handbook of management and organization inquiry*. Emerald Publishing Limited.

Heaphy, B. (2007). *Late modernity and social change: Reconstructing social and personal life*. Routledge.

Hopper, R. D. (1950). The revolutionary process: A frame of reference for the study of revolutionary movements. *Social Forces, 28*, 270–279.

Hsu, E. L. (2015). The slow food movement and time shortage: Beyond the dichotomy of fast or slow. *Journal of Sociology, 51*(3), 628–642.

Huber, G. P. (1984). The nature and design of post-industrial organizations. *Management Science, 30*(8), 928–951.

Imhonopi, D., Onifade, C. A., & Urim, U. M. (2013). Collective behaviour and social movements: A conceptual review. *Research on Humanities and Social Sciences, 3*(10), 76–85.

Inglehart, R. (1990). Values, ideology, and cognitive mobilization in new social movements. In R. J. Dalton & M. Keuchler (Eds.), *Challenging the political order: New social and political movements in western democracies* (pp. 43–66). New York: Oxford University Press.

Jenkins, J. C. (1983). Resource mobilization theory and the study of social movements. *Annual Review of Sociology, 9*(1), 527–553.

Jenkins, J. C., & Perrow, C. (1977). Insurgency of the powerless: Farm worker movements (1946-1972). *American Sociological Review, 42*, 249–268.
Jenness, V. (1995). Social movement growth, domain expansion, and framing processes: The gay/lesbian movement and violence against gays and lesbians as a social problem. *Social Problems, 42*(1), 145–170.
Jessop, B., Moulaert, F., Hulgård, L., & Hamdouch, A. (2013). Social innovation research: a new stage in innovation analysis. *The International Handbook on Social Innovation: Collective Action, Social Learning and Transdisciplinary Research*, 110–130.
Johnston, H., & Klandermans, B. (1995). The cultural analysis of social movements. *Social Movements and Culture, 4*, 3–24.
Kaltefleiter, W., & Pfaltzgraff, R. L. (1985). Towards a comparative analysis of the peace movements. *The Peace Movements in Europe and the United States* (New York: St. Martin's, 1985), 194.
Karpov, A. O. (2017). The problem of separating the notions of "knowledge" and "information" in the knowledge society and its education. *Procedia-Social and Behavioral Sciences, 237,* 804–810.
Kauffman, L. A. (1990). The anti-politics of identity. *Socialist Review, 1*, 67–80.
Klandermans, B. (1994). Review essay: Targeting the critical mass. *Social Psychology Quarterly, 57*(4), 368–372.
Kniffin, L. E., & Patterson, R. M. (2019). Re-imagining community leadership development in the post-industrial era. *Journal of Leadership Education, 18*(4).
Knottnerus, J. D. (1983). The Melanesian cargo cults: A test of the value-added theory of collective behavior. *Sociological Inquiry, 53*(4), 389–403.
Kornhauser, W. (1959). Power and participation in the local community. *Health Education Monographs, 1*(6), 28–40.
Kriesi, H. (Ed.). (1995). *New social movements in Western Europe: A comparative analysis* (Vol. 5). U of Minnesota Press.
Kunst, J. R., & Obaidi, M. (2020). Understanding violent extremism in the 21st century: The (re) emerging role of relative deprivation. *Current Opinion in Psychology*.
Larana, E. (2009). *New social movements: From ideology to identity*. Temple University Press.
Lentin, A. (1999). Structure, strategy, sustainability: What future for new social movement theory? *Sociological Research Online, 4*(3), 193–205.
Leong, C., Faik, I., Tan, F. T., Tan, B., & Khoo, Y. H. (2020). Digital organizing of a global social movement: From connective to collective action. *Information and Organization, 30*(4), 100324.
Lofland, J. (2017). *Protest: Studies of collective behaviour and social movements*. Routledge.

Loures, L., Panagopoulos, T., & Burley, J. B. (2016). Assessing user preferences on post-industrial redevelopment. *Environment and Planning B: Planning and Design, 43*(5), 871–892.

Lowe, P., & Goyder, J. (1983). *Environmental groups in politics*.

Lowe, P., Phillipson, J., Proctor, A., & Gkartzios, M. (2019). Expertise in rural development: A conceptual and empirical analysis. *World Development, 116*, 28–37.

Macionis, J. J. (2001). *Sociology* (8th ed.). Upper Saddle River, NJ: Prentice Hall.

McAdam, D. (1988). *Freedom summer*. New York: Oxford University Press.

McAdam, D., & Tarrow, S. (2018). The political context of social movements. *The Wiley Blackwell Companion to Social Movements*, 17–42.

McMichael, P. (2016). *Development and social change: A global perspective*. Sage.

McKeon, N. (2013). 'One does not sell the land upon which the people walk': Land grabbing, transnational rural social movements, and global governance. *Globalizations, 10*(1), 105–122.

Melluci, A. (1995). The process of collective identity. In H. Johnston & B. Klandermans (Eds.), *Social movements and culture* (pp. 41–63). Minneapolis, MN: University of Minnesota Press.

Melucci, A. (1980). The new social movements: A theoretical approach. *Social Science Information, 19*(2), 199–226.

Melucci, A. (1994). A strange kind of newness: What's "new" in new social movements. *New Social Movements: From Ideology to Identity*, 103.

Meyer, D. S. (2012). *Protest and political process. The Wiley Blackwell Companion to political sociology* (pp. 397–407). London: Blackwell.

Miller, F. D. (1999). The end of SDS and the emergence of weatherman: Demise through success. In J. Freeman & V. Johnson (Eds.), *Waves of protest: Social movements since the Sixties* (pp. 303–324). Lanham, MD: Rowman & Littlefield Publishers.

Mooney, P. H. (2000). Specifying the "rural" in social movement theory. *Polish Sociological Review*, 35–55.

Morrison, E. D. (1978). Some notes towards relative deprivation, social movements and social change. In *Collective behaviour and social movements*. Itasca, 111.

Müller-Rommel, F. (1985). Social movements and the Greens: New internal politics in Germany. *European Journal of Political Research, 13*(1), 53–67.

Murray, A., Kotha, S., & Fisher, G. (2020). Community-based resource mobilization: How entrepreneurs acquire resources from distributed non-professionals via crowdfunding. *Organization Science, 13*(4), 797–1051.

Nikolaeva, E. M., & Fedchenko, S. S. (2016). Contemporary mass society: The transformation of the Internet age. *International Journal of Humanities and Cultural Studies (IJHCS), 1*(1), 192–198. ISSN 2356-5926.

Offe, C., Wiesenthal, H., & Keane, J. (1985). *Two logics of collective action* (pp. 170–220). Cambridge: The MIT Press.

Oliver, P., & Johnston, H. (2000). What a good idea! Ideologies and frames in social movement research. *Mobilization: An International Quarterly, 5*(1), 37–54.

Pichardo, N. A. (1997). New social movements: A critical review. *Annual Review of Sociology, 23*(1), 411–430.

Power, S. A., Madsen, T., & Morton, T. A. (2020). Relative deprivation and revolt: current and future directions. *Current Opinion in Psychology, 35*, 119–124.

Ray, D., & Tarafdar, M. (2017). *How does twitter influence a social movement?* https://aisel.aisnet.org/ecis2017_rip/60/.

Roggeband, C., & Klandermans, B. (Eds.). (2017). *Handbook of social movements across disciplines*. Springer.

Rojas, F. (2006). Social movement tactics, organizational change and the spread of African-American studies. *Social Forces, 84*(4), 2147–2166.

Rudig, W. (1988). Peace and ecology movements in Western Europe. *West European Politics, 11*(1), 26–39.

Ryan, B. (2013). *Feminism and the women's movement: Dynamics of change in social movement ideology and activism*. Routledge.

Sawyers, T. M., & Meyers, D. M. (1999). Missed opportunities: Social movement abeyance and public policy. *Social Problems, 46*(2), 187–206.

Schneiberg, M., & Lounsbury, M. (2017). Social movements and the dynamics of institutions and organizations. *The Sage handbook of organizational institutionalism*, 281–310.

Schroeder, R. (2018). *Social theory after the Internet: Media, technology and globalization* (p. 210). UCL Press.

Shackman, G., Ya-Lin, L., & Wang, X. (2002). *Why does a society develop the way it does*. http://gsociology.icaap.org/report/summary2.htm.

Shaiko, R. G. (1993). Greenpeace USA: Something old, new, borrowed. *The Annals of the American Academy of Political and Social Science, 528*(1), 88–100.

Smithey, L. A. (2009). Social movement strategy, tactics, and collective identity 1. *Sociology Compass, 3*(4), 658–671.

Smelser, N. (1963). *Theory of collective behaviour*. Nova Iorque.

Smelser, N. J. (2011). *Theory of collective behavior*. Quid Pro Books.

Social Forces 28 (3), 270-280. Retrieved May 12, 2008 from EBSCO Online Database SocINDEX http://searchebscohost.com/login.aspx?direct=true&db=sih&AN=13582480&site=ehost-live.

Söderberg, J. (2013). Determining social change: The role of technological determinism in the collective action framing of hackers. *New Media & Society, 15*(8), 1277–1293.

Stephan, U., Patterson, M., Kelly, C., & Mair, J. (2016). Organizations driving positive social change: A review and an integrative framework of change processes. *Journal of Management, 42*(5), 1250–1281.

Stoecker, R. (2018). About the localized social movement. In *Handbook of community movements and local organizations in the 21st century* (pp. 211–227). Cham: Springer.

Strasser, H., & Randall, S. C. (Eds.). (1981). *An introduction to theories of social change*. London: Routledge and Kegan Paul.

Tarrow, S. (1994). The sources of social power, volume 2: The rise of classes and nation-states, 1760–1914. By Michael Mann. New York: Cambridge University Press, 1993. *American Political Science Review, 88*(4), 1031–1032.

Tilly, C., & Wood, L. J. (2015). *Social movements 1768–2012*. Routledge.

Tilly, C., Castaneda, E., & Wood, L. J. (2019). *Social movements, 1768–2018*. Routledge.

Touraine, A. (1971). *The may movement: Revolt and reform: may 1968—the student rebellion and workers' strikes—the birth of a social movement*. Random House (NY).

Touraine, A. (1981). *The voice and the eye: An analysis of social movements*.

Tremblay, M. C., Martin, D. H., Macaulay, A. C., & Pluye, P. (2017). Can we build on social movement theories to develop and improve community-based participatory research? A framework synthesis review. *American Journal of Community Psychology, 59*(3–4), 333–362.

Van Dyke, N., & Amos, B. (2017). Social movement coalitions: Formation, longevity, and success. *Sociology Compass, 11*(7), e12489.

Vasin, S. M., Gamidullaeva, L. A., & Rostovskaya, T. K. (2017). *The challenge of social innovation: Approaches and key mechanisms of development*.

Vogt, K. C. (2016). The post-industrial society: From utopia to ideology. *Work, Employment & Society, 30*(2), 366–376.

Waddock, S. (2017). *The difference makers: How social and institutional entrepreneurs created the corporate responsibility movement*. Routledge.

Walsh, E. J. (1981). Resource mobilization and citizen protest in communities around three mile Island. *Social Problems, 29*(1), 1–21.

Weeber, S. C., & Rodeheaver, D. G. (2003). Militias at the millennium: A test of Smelser's theory of collective behavior. *Sociological Quarterly, 44*(2), 181–204.

Wieviorka, M. (2005). After new social movements. *Social Movement Studies, 4*(1), 1–19.

Wilbur, A. (2013). Growing a radical ruralism: Back-to-the-land as practice and ideal. *Geography Compass, 7*(2), 149–160.

Yamada, R. (2016). Karl Mannheim on democratic interaction: Revisiting mass society theory. *Human Affairs, 26*(2), 93.

CHAPTER 3

Social Movements as Drivers of Paradigm Innovations in Rural Development

Dalia Vidickienė

3.1 Key Challenges to Rural Development Based on Post-industrial Paradigm

Since the end of the Second World War, changes in the rural economic and social structure have been dramatic. In comparison with urban inhabitants, rural population experienced more transformations in all spheres of their life during this period. The biggest challenge is facing 'baby boom' generation, born between 1946 and 1964, which must adopt the second socio-economic paradigm change during their life.

The first paradigm change started by the mass industrialization of agriculture. Although urban dwellers in the developed world have lived by the laws of industrial society for several centuries, the industrialization of the agricultural sector and rural life began not so long ago. The authors studying the economic history of agriculture claim that the mass industrialization (often referred to as 'modernization') of agriculture began only after the Second World War (Federico 2005; LaVerne 2012; Martín-Retortillo & Pinilla, 2012, 2013).

The following factors have been key drivers of the mass industrialization of agriculture:

R. Vilkė et al., *Rural Economic Developments and Social Movements*, Palgrave Advances in Bioeconomy: Economics and Policies, https://doi.org/10.1007/978-3-030-71983-8_3

- The decreased demand for defence production. The new technological knowledge, industrial equipment and human skills to produce tanks and other military vehicles and equipment have been used to design and produce innovative agricultural technique.
- A lack of food in the countries involved into active participation in the actions of the Second World War.
- A lack of working force in countryside because when the war began, many farmers and hired farm-hands enlisted in the military or migrated to war industry jobs in urban areas.

In these circumstances all efforts of national policies and innovative farmers have been focused on productivity of available working force in agricultural sector. The production destroyed by the war-induced a wide variety of labour-saving and crop yield increasing technological innovations. All innovations have been focused on the substitution of the main production factors in the agrarian society—land and labour—by artificial inputs produced by manufacturing sector. Industrialization of agriculture was based on four groups of technological innovations:

- Agricultural machinery (until the Second World War, agricultural machinery was used *en masse* only by grain growers in the Great Plains and Pacific regions of North America and the Soviet Union);
- Irrigation or melioration;
- Electrification;
- Chemical fertilizers and pesticides.

Due to the massive introduction of mechanization, electrification, irrigation and chemicalization processes after the Second World War, the agricultural sector in the developed industrial (northern) countries has undergone a real revolution (Clunies-Ross & Hildyard, 2013; Harwood, 1990). Industrialized agriculture essentially turned farms into factories, requiring inputs like synthetic fertilizers, chemical pesticides, large amounts of irrigation water and fossil fuels to produce crops and livestock by mechanized production means. The farmers began to depend on money, rather than land, to feed themselves. The production mode became the same as that applied in the industry: mechanization and segmentation of work, specialization of output, standardization of product. Agriculture had become capital intensive and farmers' capital and credit needs were growing at record paces. Farmers' business success had become heavily dependent on the external factors, i.e. on the situation in the manufacturing sector and the market.

The era of agricultural sector industrialization can be defined as a period which is associated with what is known as productivism with respect to rural locations as all of mentioned technological innovations became an important means changing productivity of agricultural sector. Productivism was introduced as a model that positions agriculture as a progressive and expanding food production-orientated industry (Marsden et al., 1993). A lack of food was a strong driver for expanding amounts of food production and a lack of working force-driven implementation of technological innovations. Industrialization opened new opportunities for productivity gains as new equipment enabled even fewer farmers to perform their work in a more timely fashion than had previously been possible. After the introduction of new agricultural technology, productivity increases have been very rapid. The average annual growth rate in 1950–1992 was even 4.46% (Martín-Retortillo & Pinilla, 2012, p. 6).

By the 1970s, however, negative effects of industrial production mode became visible. Increased use of machines and synthetic fabrics strained the farming business, subjecting it to a greater need for capital and chronic overproduction, which lead to low prices of agricultural products. After experiencing problems of surplus, farmers became increasingly dependent upon exports as a market for their plentiful production. The period of overproduction was characterized by unprecedented changes in how food is produced, distributed, consumed and controlled—and by high levels of concentration of market share. Organizational and institutional change in the relation of farmers with customer and farm input and processing companies happened as the agri-food supply chain, which is a sequence ranging from inputs to the final delivery, absorbed more and more different actors. Many intermediaries as processors, warehouses, transporters and traders stepped between the farmer and end users of their food products. Increasing number of intermediaries in the food supply chain had a profound impact on farmers' incomes. As a result, farmer net income has been declining over multiple decades as the prices that producers are receiving at the farm gate for their commodities are not rising at a rate equivalent to the price of their inputs. If in the sixties and seventies farmers received 40–50% of the food prices set by supermarkets, the share of farmers in developed countries since the eighth decade of the twentieth century is usually not more than 10% (Guthrie et al., 2006).

The more an agricultural sector was managed to provide maximum yield the more vulnerable it was to external shocks. In the late twentieth century, the problems of industrial production mode gained added complexity. The following economic problems were particularly notable: rising levels of capital intensiveness in farm production, rising costs of inputs, declining commodity prices and food market concentration in the processing and retail sectors. In parallel with economic problems, a variety of negative social and environmental effects arise in countryside. Rural communities have undergone a fundamental social restructuring as industrialization gave rise to large agribusinesses and new farmers' elite while worsening prospects of small farmers. Since economic power was unevenly distributed, the chances of satisfying the needs for countryside values for rural population became unequal. In such conditions, the competition for space often results in spatial segregation of the population according to wealth, fragmentation of rural space and locally increasing motor traffic (Frouws, 1998). Alongside economic and social problems has come an increasing international awareness of issues of environmental degradation caused by agriculture. The authors analysing the drivers of change in global agriculture emphasize several environmental concerns including deforestation and forest degradation; water depletion and degradation of irrigated land; soil degradation; biodiversity losses; global and regional climate change (Hazell & Wood, 2008).

The agroindustrialization also was a process stimulating a dramatic rural-to-urban movement and leading the countryside to a distinctive economic and social system exhibiting the characteristics similar to urban. A global-scale demographic shift, known as depeasantization, began as the rural agricultural population moved to urban areas (Araghi, 1995). The post-Second World War period, was the first period in which the peasantry became a minority, not merely in industrialized developed countries, in several of which it had remained very strong, but even in the Third World countries' (Hobsbawm, 1992, p. 56). Whereas 29% of the global population, and 16% of the 'Third World', lived in urban areas in 1945, by 2012 that number had raised to 51 and 47%, respectively (World Bank, 2013).

At the beginning of twenty-first century intensive industrialization of agriculture has increasingly been critisized and suggested to be abandoned in an effort to reduce the negative outcomes associated with this production mode. There is a growing body of academic literature that argues that the agro-industrial food security model that prevailed after the Second World War has experienced a crisis and needs to be improved (e.g. Marsden, 2003; Nemes, 2005; Rosin et al., 2013; Van der Ploeg et al., 2010; Vidickiene & Melnikiene, 2014; Weatherell et al., 2003; Wolf & Bonanno, 2013). During the last decades two discourses on the future of the agricultural and food systems and countryside can be found. The discourses can be defined as a group of ideas or patterned way of thinking based on the belief that the current rural development strategy should be: (1) revised or (2) fundamentally changed.

The first group of researchers and politics believe that agro-industrial system is the best and the only way forward for human development. They are developing a discourse which is trying to find an answer to the question: 'How to overcome negative effects of the agro-industrial system while retaining the same basic structure and ways of functioning?' This group is advocating the 'productivity narrative' which is based on the mental model that reflects the industrial paradigm. The main assumption is that scientific advances have the potential to bring forward new technologies, varieties and breeds that could boost productivity and at the same time to take into account resource scarcities and environmental problems. In their opinion, all negative outcomes of the industrial agrotechnologies can be improved by the creation and implementation of new technological innovations.

The proponents of industrial development are looking for ways to avoid or mitigate negative effects of the industrial mode of agriculture emerging at the maturity stage of its life cycle through the framework of sustainable development. The concept of sustainable development is focusing on balancing environmental, economic and social needs by socially and conservation-conscious approaches. Based on the belief that agro-industrial system has not alternatives, a huge share of the public financial resources flows to maintain the sustainability of industrial agriculture mode. Many rural social movements support this idea, demanding ever-increasing financial assistance for farmers by a number of supportive measures that should be taken by the governments. However, current approaches to sustainability, focusing largely on efficiency and productivity improvements and 'greening' supply chains and products usually do not go beyond the industrial paradigm.

According to the proponents of post-industrial paradigm, the feasibility of pursuing the industrial way of economic development should be brought into question. Advocates of the second discourse believe that further evolution of agriculture and countryside is possible only with the introduction of paradigm changes based on the theory of post-industrialism. Following earlier agrarian and industrial 'revolutions', post-industrialism suggested yet another revolution that would again transform how societies were organized (Hoey, 2015). 'There is a need for a new rural development paradigm that can help clarify how new resource bases are created, how the irrelevant is turned into a value and how, after combining with other resources, the newly emerging whole orientates to new needs, perspectives and interests' (Van der Ploeg et al., 2000, p. 399).

The new rural development paradigm is emerging as a set of responses to the old, agroindustrialization paradigm by several paradigm innovations ('paradigm innovations are changes in the underlying mental models which frame what the organization does') (Bessant & Tidd, 2007, p. 13). Not all of them are well understood. We are seeing clear signs of the paradigm nude, however, do not know how to reorganize our life in the new reality. The major challenges for development policymakers dealing with differences between industrial and service-oriented post-industrial socio-economic system are covering following questions: (1) what kind of innovations are driving the socio-economic development in the service economy, (2) how successful business model must be designed, and (3) how economic and social relations should be reorganized in order to generate synergetic effects. Each question helps to take advantage of the new opportunities for rural regions offered by the post-industrial economy and clearly define three key paradigm innovations:

1. Turning from technological to non-technological drivers of development;
2. Transitioning from product-driven to service-driven business model;
3. Shifting from competitive or exploitative to collaborative and synergistic relationships.

The role of social movements in rural development is also changing in the context of the paradigm innovations. Examination of post-industrial social movements through the prism of mentioned three paradigm innovations opens new ways of thinking and doing. For taking three paradigm innovations as theoretical background in post-industrial social movement studies it is necessary to apply transdisciplinary approach to the perception of the mentioned paradigm innovations.

3.1.1 Paradigm Innovation 1. Focusing on Non-technological Drivers of Development

In the industrial era term 'innovation' was commonly associated with progress through technological modernization and economic growth (Blok & Lemmens, 2015; Grunwald, 2017; Ludwig & Macnaghten, 2020; Scott-Smith, 2016; Wells, 2018). The proponents of radical changes in rural development paradigm emphasize a role of social innovations (Bosworth et al., 2016; Katonáné Kovács et al., 2016; Navarro et al., 2018; Neumeier, 2017). Some authors even claim that we are living under the social innovation imperative (Harris & Albury, 2009). As point out Edwards-Schachter and Wallace (2017), after decades of marginalization, social innovation is starting to be recognized as part of the 'black box' of innovation to inquire which is the 'place' of social practices in innovation processes and how they take part of different activities and the building of social, technological even cultural innovation systems and their contribution to sociotechnical change.

The need to study innovation beyond industrialization based Western paradigms, also confirms experiences from emerging economies, such as China, India and Latin America (Chen et al., 2018) that are focused on another feature of post-industrial drivers of development. Describing 'networks of activists and organizations generating novel bottom-up solutions' Seyfang and Smith (2007, p. 585) introduce the term 'grassroots innovations'. This term is often used for the analysis of nature, goals and benefits of innovations for rural development in developing countries (Kirwan et al., 2013; Seyfang & Haxeltine, 2012; Seyfang & Smith, 2007; Smith et al., 2016). Grassroots innovations differ from top-down solutions as they involve in innovation design people at the community level. They are not top-down in nature but instead emerge from the bottom-up as a result of pilot projects developing alternatives to the mainstream hegemonic regime. To make local innovative solutions widely viable at grassroots level the local initiatives and pilot projects often are coordinated by grassroots innovation movements. Through the empowerment, capacity-building and development of raised levels of awareness, communities and grassroots innovation movements have the potential to make a contribution to more profound 'paradigm change' within society (Seyfang & Smith, 2007; Smith et al., 2013).

The latest innovation studies are based on the service economy features and take a more integrative approach to the classification of innovations. In the industrial era, innovations have been classified as product, process, organizational, marketing, management and social innovations. Currently more and more often innovations are divided into technological and non-technological innovations. This classification is gaining popularity based on the intuition that the industrialization paradigm is driven by technological innovations, but the post-industrial paradigm is driven by the non-technological innovations. Most of the non-technological innovations are complex in nature and can join together several or all possible types of innovations usual for the industrial era including service (product), process, organizational, marketing, management and social innovations.

Rural development paradigm also should turn the focus from technological innovations to non-technological innovations covering various organizational aspects of economic, social and ecological rural systems. At the same time, we should keep in mind that the evolution does not ignore the role of previous important factors in the development process including technological innovations. Focusing on non-technological innovations doesn't have to mean that technological innovations will not be used as a tool for rural development.

3.1.2 Paradigm Innovation 2. Transition to Service-Driven Business Model

Since the 1990s, the management research has made a step forward in understanding the complex nature of post-industrial non-technological innovations by evolving concepts of 'business model' and 'business model innovation'. Business model innovation is a concept which is blurring lines between different types of non-technological innovations and explains how they are working as a unified whole. The paradigm business model innovation is servitization which includes innovations in strategy, marketing, management, and organizational process, supply chains, pricing and cost structures. Over the last couple of decades, the term 'servitization' has become one of the most popular new terms describing the key paradigm innovation of 'new economy'. Servitization has been extensively studied as a post-industrial way of planning and doing business (Baines et al., 2017) and as a key tool for regional development. Research on territorial servitization issues show that servitization creates new local productive configurations (Bellandi & Santini, 2019), strengthens territorial competitiveness (Vendrell-Herrero & Wilson, 2017; Gomes et al., 2019), regions also benefit from servitization processes via the interplay of generating employment opportunities, enabling an efficient allocation of technology and resources, opening up new markets, raising the odds of securing employment in the consolidation period and enabling technological leaps (Gebauer & Binz, 2019). At the current state of the art the servitization can be defined as 'the penetration of service delivery elements into all areas of the economy by the gradual shift from product-driven business model to service-driven business model' (Vidickiene, 2017, p. 474) and this penetration is similar to penetration of industrial innovative technologies at the beginning of industrialization era.

The scholars still are concentrated on transition to service-driven business model in manufacturing. However, the servitization can also be an important tool for changes in rural development as service-driven business models consider how value may be fashioned and realized via more dynamic, interactive arrangements between consumers and producers. In most cases the servitization of farming and other rural businesses can be the most progressive way to respond to major rural development challenges, especially for generating desirable changes in industrial food supply chains as service-driven farming business model reduces food demand and waste and provokes desirable behavioural changes of consumers. Although service-driven business models are already used quite often in the farming practices, they usually are not analysed in the context of servitization and the concepts of the industrial economy are still used while analysing innovations in the agricultural sector dealing with the servitization of farming (Vidickiene & Gedminaite-Raudone, 2018). For example, all efforts made by farmers to apply new business models to respond to the needs of consumers for fresh, locally produced food are defined as the desire to shorten food supply chain. However, the farmers are using different business models and the business model when a farmer aims to sell directly to the consumer in order to reduce the number of intermediaries involved in the supply chain can be fundamentally different from the service-driven business model when a farmer aims benefit by providing more services.

3.1.3 Paradigm Innovation 3. Building of Collaborative and Synergetic Effect Generating Relationships

Competition is a central element of industrial product-driven industrial society, but in an extremely high competitive environment competition is a zero-sum game when we want to win at the expense of the other persons. In other words, the success of one side is predicated on the failure of the other side and the relationships do not generate synergetic effect. This situation is taken for granted because the product-driven industrial economy is based on the bilateral economic exchanges and price negotiations do not suppose synergetic effect, but on the contrary both sides are seeking a 'bigger piece of the pie'.

The evolution of the wildlife shows that conflicts between organisms have resolved into symbiotic partnerships. The same path of the evolution goes to the human society shifting focus from competitive to collaborative relationships in business and other spheres of life and learning to combine both types of relationships. One can actually strengthen the other and should not be considered as mutually exclusive forces. Service-driven post-industrial economic system suggests supplementing industrial model of customer–vendor bilateral economic exchanges by multilateral collaboration model which is a key success factor in client and service provider relationships. Today's service providers are successful because they have a mindset of solving the individual clients' problems and needs, rather than merely offering standard solutions. As pointed out by Palmatier (2008), the confluence of factors, including the transition to service-based economies; advances in communication, logistics and computing technologies; increased global competition; and faster product commoditization have enhanced the salience of 'relationship-based loyalty' to sellers compared with other marketing mix factors. In result, relationship marketing and customer relationship management have taken a central position in marketing strategy in the twenty-first century.

Building collaborative capabilities becomes of paramount importance in servitizing rural regions. To succeed, the rural entrepreneurs as service providers should constantly balance both moving forward with daily tasks and demonstrating their value to the client in order to build mutually beneficial symbiotic relationships and generate synergetic effect. Value creation is no longer perceived to reside within farm or rural enterprise boundaries but the value is considered to be co-created between various actors within the networked structures. This development poses major challenges in all types of relationship-building for rural businesses and development policymakers.

All three discussed paradigm innovations serve as a three dimensional framework for specific place-based approaches (Fig. 3.1). The first dimension—focus on non-technological drivers of progress—frames the space of innovative activities. The second dimension—a shift from product-driven to service-driven business model—is desirable in all innovative solutions. The third dimension—a shift from competitive or exploitative to collaborative and synergistic relationships—is the organizational construction. It is used as a ground for all changes required for the post-industrial rural development paradigm building.

The framework shows only the direction of desirable radical changes. The implementation of the paradigm innovations requires a lot of place-based creative solutions responding effectively to certain business or community challenges. The next chapter discusses who can help the rural entrepreneurs and communities to overcome or break down barriers to key paradigm innovations in rural development in the most effective way.

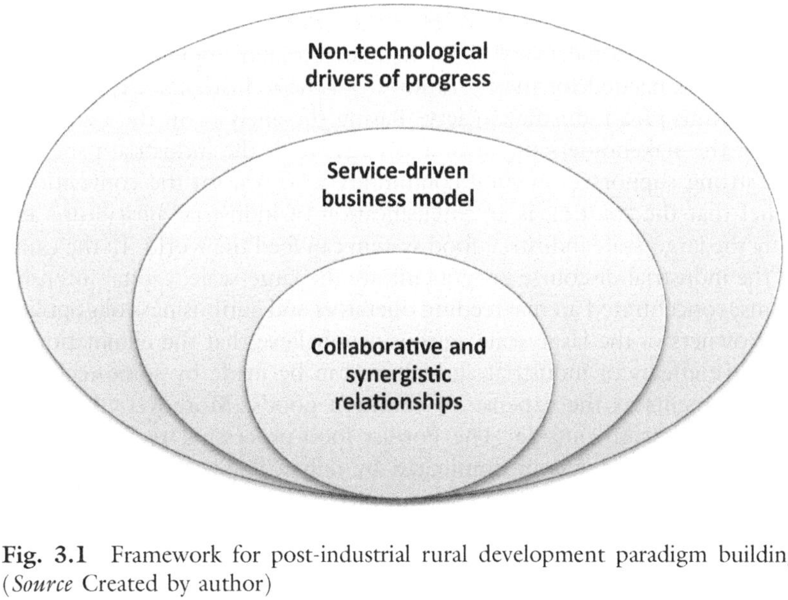

Fig. 3.1 Framework for post-industrial rural development paradigm building (*Source* Created by author)

3.2 New Stakeholders and Goals of Social Movements for Rural Development at the Beginning of the Twenty-First Century

Diffusion of any innovation depends on efforts of stakeholders. Whether the post-industrial paradigm has enough supporters who can implement the necessary innovations? What actors have the biggest potential to design and implement three mentioned key paradigm innovations and act as drivers of rural development in post-industrial society?

The revitalization of the countryside according to the new rural development paradigm was primarily conceptualized as an endogenous process (Van der Ploeg & Long, 1994; Van der Ploeg & van Dijk, 1995), depending to a great extent on local resources. This suggestion altered the theoretical perspectives on place-based rural development. OECD (Organization for Economic Cooperation and Development) started to advocate place-based rural development as New Rural Paradigm (OECD, 2006).

However, the needed radical changes in the socio-economic systems raise questions about the potential of rural population to reorganize countryside according to the new rules of the game in post-industrial society. Are the rural people ready alone to generate and implement paradigm innovations needed for the transition from the industrial socio-economic system into post-industrial society? Basing the analysis on the traditional list of the stakeholders the answer is negative as the industrial paradigm has strong supporters in rural community who rest on the conventional belief that the solution is an intensification of industrial agriculture and only the large-scale industrial food system can feed the world. In the camp of the industrial discourse proponents are the large-scale, capital-intensive farms, concentrated animal feeding operators and agribusiness oligopolies. The owners of the large-scale agribusiness believe that the elimination of negative effects of industrial agriculture can be made by some technical improvements at the expense of common goods. Moreover, they have the support from outside. The Fordist food-processing techniques had resulted in a food system dominated by only a few companies sourcing foods from all over the world (Adams, 2016) and an economic and spatial power is concentrated in the hands of the leading food manufacturers and retailers. The power of large-scale retailers and agribusinesses best illustrates the case how they took the lead in transforming food systems in Central and Eastern Europe when the transition had happened from state-organized economic systems to market-based systems (e.g. Barrett et al., 2019; Reardon et al., 2003; Rozelle & Swinnen, 2004).

In the camp of the post-industrial discourse proponents among rural populations are mainly small farmers and rural entrepreneurs. The small farmers usually are passive players. The majority among them is unorganized and lacks political experience to challenge existing status quo forces. They are not considered to have the capacity to assume political power in the absence of concerted efforts to get them involved and organized.

Since economic power of rural stakeholders in rural development is unevenly distributed, looks like post-industrial discourse proponents have little chance of winning. Success in rural development paradigm shift requires mobilized organizations of rural population able to channel demands and ensure that the reform process meets the needs of society. The doubts grow about the socio-economic strength of the rural community in meeting social demands not only for high-quality authentic food and wholesomeness, but also for the building of other desirable countryside features as in the frame of industrial paradigm dominates the 'sectoral' approach to multifunctionality. Many works dealing with the non-market functions of agriculture have not been included within the framework of multifunctionality of rural development (Sumelius & Bäckman, 2008). However, in the context of relations between agriculture and society, a 'wider' approach to rural development emerged in the 1990s (Torre & Wallet, 2016). The rural is no longer perceived as the exclusive domain of agriculture and part of the growing public concern for the countryside stems from worries about the loss of qualities like open space, peace and quiet, scenic beauty, old man-made landscapes, biodiversity and pleasant villages (Frouws, 1998). However, mental models are enduring and resistant to change (Gentner & Stevens, 1983). The powerful key players representing agribusiness and large-scale retailers are not ready for radical change in food production mode and the spatial scaling of everyday foodways. They are also against the proposals part of the agricultural land turn into nature reserves, recreation parks and green residential areas. They tend to ignore the destructive role of intensive agriculture and push policymakers to keep the same development trajectory. It seems that the task to rebalance industrial paradigm based thinking is too difficult for current institutional structure.

However, the new look to social basis opens new interest groups ready for post-industrial changes. Till now the rural development stakeholders typically included actors along the food supply chain, with farmers and land managers at the forefront. The variety of competing and also frequently conflicting claims and functions attached to the countryside is causing the extension of the rural development stakeholders list formed in the industrial era. Diffusion of each paradigm innovation is introducing requirements to a new social basis of rural development by creating opportunities to involve new stakeholders and decreasing the role of stakeholders formed in the industrial era.

The first paradigm innovation requires the involvement of new stakeholders by shifting focus on non-technological innovations as key drivers of progress. The key challenge to rural development stakeholders creates intangible character of non-technological innovations. Most of the technological innovations used for the industrialization of agriculture were tangible. A key task of farmers was to find capital for new agricultural machinery, equipment, synthetic fabrics and other technical arrangements and learn to use them. Post-industrial paradigm represents a significant shift from investment in physical capital and engineering knowledge to investment in soft assets needed to run service business by developing the communication, marketing, project management and other knowledge and skills. In many cases, the required investments are rather small in comparison with implementation of industrial paradigm. The key resources are the creativity and ability to take multidisciplinary approach for business knowledge as the innovation has shifted from being engineering-driven to design-driven and from marketing-focused to user-experience-focused. The differences between non-technological and technological innovations give a power for stakeholders with different features. The intangible nature of key resources gives more strength for people with creative thinking and organizational skills in contrast to success factors of the industrial era: memory-based learning and standard operating. In such circumstances, the role of leading stakeholders is taking by socially oriented entrepreneurs, including farmers, and NVOs with high power to implement non-technological innovations.

The second paradigm innovation also requires new stakeholders as it becomes clear that smooth transition from an industrial to a post-industrial service-driven business model is often impossible only through business efforts. As the business model innovations are mainly intangible inventions, the innovation creation is very different process in comparison with the creation of technological process or tangible product innovations. Very few service firms rely on traditional R&D with regard to their innovation activities (Miles, 2008). Rather than starting with

R&D activities, as there is typical for industrial era, implementation of new services or rebuilding of old offers usually starts with idea screening, service design and service testing in close cooperation with potential customers (Song et al., 2009). In industrial society, innovation has been thought of as a producer-centred process based on the assumption that profit-seeking incentives are the main driver of innovation (Trischler et al., 2020). However, in post-industrial society users became more and more important source of innovation and they are innovating for-use rather than for-sale (Baldwin & von Hippel, 2011). Post-industrial service-driven paradigm integrates supply-oriented and demand-oriented approaches. The last rural innovation studies emphasize the multi-stakeholder approach, but most of them still are based on the supply-oriented industrial paradigm and compile a list of stakeholders generating economic values according to supply chain. The growing literature on new ways of rural development emphasizes a need for user-driven innovation (Arabska et al., 2014; Guzmán et al., 2013; Zavratnik et al., 2019). With user-driven innovation, the scholars refer to an activity conducted by any type of users (e.g. individual consumers, firms and NGOs) who spend their unpaid discretionary time developing innovative solutions to address their personal needs (Edwards-Schachter & Wallace, 2017; Gambardella et al., 2017). As a result of the shift from product-driven business model to service-driven business model, no more exist the generation of 'pure' economic or social values. The abundant academic literature confirms that the generation of 'blended' values blurring boundaries between profit and non-profit sectors (Borzaga & Bodini, 2014; Edwards-Schachter & Wallace, 2017; Mulgan et al., 2007; Pol & Ville, 2009). Implementation of service-driven business model requires an innovative combination of activities across a wide range of economic sectors and subsectors and provides a new rationale for having territorially based trans-sectoral innovation systems (local, regional or national). This feature of the service-driven business model is a call for new activities of social movements as drivers and active co-creators of local, regional, national and international innovation systems.

This trend also impacts the structure of rural development stakeholders. According to Lundvall (2013, p. 33), 'the closest we get to such a core in innovation studies is the conceptualization of innovation as an interactive process involving many actors and extending over time'. Transformation of a set of actors involved in rural development firstly requires that the active, relational and political role of consumers in the genesis and reproduction of these new economic forms be 'acknowledged' (Goodman, 2004). To imagine the radical change in food production, systems of provision and the spatial scaling of everyday foodways without the agency of consumers is simply quixotic, given the formidable economic and spatial power concentrated in the hands of the leading food manufacturers and retailers (Heffernan et al., 1999; Hendrickson & Hefferman, 2002).

Service users as sources for innovation has an especial role (Skiba, 2010). The latest management theories are emphasizing the role of co-creation in service provision, where a part of the new value is generated not by service providers but by the clients (Ramaswamy & Ozcan, 2014). For instance, when the farmers implement the business model 'product plus service' instead of supplying their products to an anonymous food market, they need to create long-time collaborative relationships with clients (Vidickiene et al., 2019). The research provides evidence that customers also are willing to develop social relationships with service providers (Gittell, 2002; Suhartanto et al., 2019). The degree of inclusion of a service consumer in a newly created value can be varied, but the clients always make their own contribution. Considering this fundamental change in the value creation process, the key future actors in rural development will be service users as core actors in service-driven business model building process.

Analysing new rural development stakeholders should be taken into account not only individual economic actors but also social collectives as various NGOs and informal communities dealing with rural development innovations. Their number and influence grow as the internet is, therefore, now a major starting point of new social formations. The Internet has encouraged emergence of online communities with a special thematic focus. The scholars find different types of them (Dolata & Schrape, 2016), including epistemic communities, defined as a network of professionals with recognized expertise and competence in a particular domain (Haas, 1992); communities of practice, whose participants deal with similar (professional) tasks (Wenger, 1998); brand communities, who share a sense of togetherness around a brand (Fournier & Lee, 2009), etc. Social media enables ordinary citizens to connect and organize themselves with little to no costs, and the world to bear witness. The formation of new collective actors dealing with rural development increasingly occurs through online-based communication. They often start with little more than unstructured collective behaviour; many of them eventually turn into organized forms of collective action. The role of informal collectives as communities of interest in rural development issues dealing with post-industrial paradigm especially grows in information exchange for problem-solving and co-creation actions as a major part of the process of non-technological innovation.

Understanding of the true value of new stakeholders' involvement calls to mobilization of non-rural population as active participants, particularly in national and international levels of the rural development. Today many active actors of social movements for rural development are urban inhabitants or newcomers from the city because the space into which the paradigm innovations should be introduced covers not only rural areas. In many cases, urban inhabitants are the leaders of social movements for rural development. Understandably, they operate through the rural people by involving rural population into processes of co-creation and diffusion of social, economic and environmental innovations. In order to encourage their activity and social impact, there is a need to create new tools and systems that encourage and support collaboration between rural and non-rural stakeholders.

The third paradigm innovation requires skills of network management. The rural post-industrial non-technological innovations should be built by networking and communication among different actors. Collaboration between stakeholders with heterogeneous skills and capabilities, particularly with new ones, is helpful for capturing more opportunities for rural development. At the same time there is a challenging process as, during this process, they should establish 'relationships'. The relationships between new players in innovation game very differ from industrial relationships based on the exchange logic. The typical exchange relations have bilateral nature and are formed in a linear way as relations between actors of supply chain. But in service business pre-dominate non-linear relationships. The disparate elements of a service ecosystem make more sense when they are viewed as synergistic parts of a whole, especially concerning value (Meynhardt et al., 2016). The establishing of post-industrial relations is based on non-linear logic and may be interpreted as forming networking organizations. The key goal of relationships building is co-creation resulting in the synergetic effect. To succeed, new tools are needed for coordination and management of multifaced interests of heterogeneous stakeholders.

The success in generating synergetic effects highly depends on the complementarity of the network members' capabilities and on the ability to combine them in order to guarantee the success of a whole network. Most innovations are now being created and implemented by multiside and platform-based networking structures (Boudreau & Hagiu, 2009; Chang-seok et al., 2020; Kim, 2014). The innovation through open participatory platforms favours holistic approach which enables:

- Joint work and harmonization of different and convergent interests of all possible stakeholders;
- Cross-sector innovations design;
- Innovation diffusion acceleration.

The most fascinating ability of this new type of networking is that innovation platforms are an ideal tool for building and development of post-industrial social movements. Changes in rural development policy paradigm goes very slowly, there are many factors that stop this process. People are realizing that the large difference between the present situation and how it should be according to the public needs may be decreased only by mass social movements because where there are choices between conflicting interests, the scope of the social basis is very important. More and more people are involved by multiside innovation platforms for taking part in alternative, synergetic effects generating initiatives of rural development. The rise of multiside and platform-based networking structures are particularly useful for involving into social movements rural people from remote regions. The movements maintain internal cohesion primarily through the formation of communication platform for the development of specific innovation. The platforms are powerful mobilizing devices for recruiting and retaining members, initiating and coordinating activities, and acquiring resources.

The changes in the set of stakeholders for rural development are pushing social movements to widen the scope of their activities in comparison with social movements built at the industrial stage of society's evolution. The industrial stage has been devoted for the developmental welfare state-building (Daly & Rake, 2003; Lewis, 1992). The coordination of different policies for economic growth and the developmental (welfare) state was typically associated with bureaucratic dominance in policymaking (Goodman & Peng, 1996; Holliday, 2000). The mission of rural social movements who emerged in the industrial era is the involvement of rural people in active protest actions against regulatory decisions based on the top-down approach that are reducing farmers' income level or economic potential by redistribution of agricultural land, various restrictions of economic activity and inadequate financial support. They are focused on changing political decisions by redistribution of profit, land, wealth and social power on behalf of farmers and other rural population.

The post-industrial social movements have more ambitious goals. The first distinctive feature of post-industrial social movements for rural development is that instead of claiming the produced value redistribution they refer to value creation and are involved in experimentation with practical alternatives to industrial countryside and agri-foods system. In contrast to rural movements of industrial era, focused on the fair sharing of value between rural and urban population, they are looking for new ways how to create new value beneficial to the society as a whole. Evolved from the industrial social movements, whose essential feature is mobilization of heterogeneous stakeholders for collective protest, post-industrial social movements have assumed a number of innovation functions formerly monopolized by business organizations. At the beginning of twenty-first century they become key laboratories for testing innovative business models and alternative means of socialization. Their innovative behaviour also is an alternative to government-led development interventions. The post-industrial social movements save public money by support to rural entrepreneurs and communities to overcome or break down barriers to paradigm innovations.

Another distinctive feature of post-industrial social movements for rural development is that they help to design and implement innovations important and beneficial not only to the rural people but to society as a whole. They emerge in a context of increasing pessimism about achieving desirable change through protests against rural development policy decisions in the frame of the industrial system-oriented institutional structure. An important factor also is sceptical or sometimes hostile attitude to efforts and ways of farmers to influence agricultural and rural development policy. The farmers are increasingly blamed for destroying nature by polluting the environment and taking a significant part of national budgets for their own welfare. The countryside is multifunctional by its nature, but at the beginning of the industrial era it was tempting to think of rural areas as merely providing food and other products such as timber or minerals. Diversification of rural employment structure away from agriculture at the last stages of industrial paradigm life cycle opened new opportunities to rural development. The role of the countryside arguably becomes more complex by taking into account the needs of the entire population of the country and considering rural regions as a space to live, work, recreate and travel which is important for everybody. In this context, 'rural development' became more sociological than geographical concept. The key challenge is to find innovative ways for rural development on behalf of all society while maintaining their distinctive rural character. In this changed environment the area of interest and aims of social movements are changing also. In post-industrial era, 'rural social movements' are transforming into 'social movements for rural development' as the future of the countryside is becoming an issue of public concern.

The next section is analysing how the global social movements for post-industrial rural development are building platforms as tools to implement new paradigm of rural development by changing relationships in all spheres of rural life.

3.3 Areas of Interest of Global Post-industrial Social Movements for Rural Development

Persistent social, economic and environmental failures of the industrial agri-food system have spurred the formation of tens of thousands of local, national and international social movements concerned with food and agriculture (Hawken, 2007). Also grows a number of social movements and organizations experimenting with alternative rural lifestyle models and taking care of rural areas as a comfortable and quite place to live or relax.

Implementation of paradigm innovations requires revision and change of a very wide set of the industrial countryside and agri-food system parameters. Each social movement for rural development begins with mobilization of stakeholders around specific issues as at the beginning of the paradigm change many aspects of the new paradigm are vague and complex approach is difficult to implement. Consequently social movements for rural development are implemented in many different forms and on different levels in order to transform industrial rural development paradigm. These social movements have involved millions of change agents and have developed a wealth of political, technical, organizational and entrepreneurial skills. Many of them have developed into powerful international organizations.

Conceptually, all post-industrial social movements are focused on the reorganization of current relationships in the context of the growing multifunctional role of countryside. Systematic analysis of key ideas on rural development through the prism of paradigm innovations framework shows that the initiatives of social movements dealing with the generation of innovations for rural development are mainly concentrated to revision and change of industrial mode of relationships. Taken together, social movements cover a wide range of societal demands to rural development that include new attitudes to our relationships with farming, nature, food and neighbours in the context of disappearing rural/urban dichotomy. Basing on the third dimension of the post-industrial rural development framework—collaborative and synergistic relationships—as the organizational construction for the activity, the social movements invent and by pilot projects implement a shift from systems based on competitive and/or exploitative relationships to systems based on synergistic relationships. This dimension is used as a ground for all changes required for the post-industrial rural development paradigm building.

What are the key types of relationships that drive transition from the industrial paradigm of rural development to post-industrial servitization based paradigm of rural development? Moving beyond single movements, and considering dynamic interactions among a multitude of contenders, following major areas of interest of well-known global social movements involved in rural development should be noted:

1. Farmers/nature relationships
2. Consumer/food relationships
3. Farming business/customers relationships
4. Rural/urban business relationships
5. Relations with neighbours
6. Peasantry/society relationships.

The revision and reorganization of all six types of relationships is a mission of global post-industrial social movements for rural development. As showed in Fig. 3.2, all types of relationships as key aspects of change in post-industrial rural development paradigm are intertwined and dependent on each other.

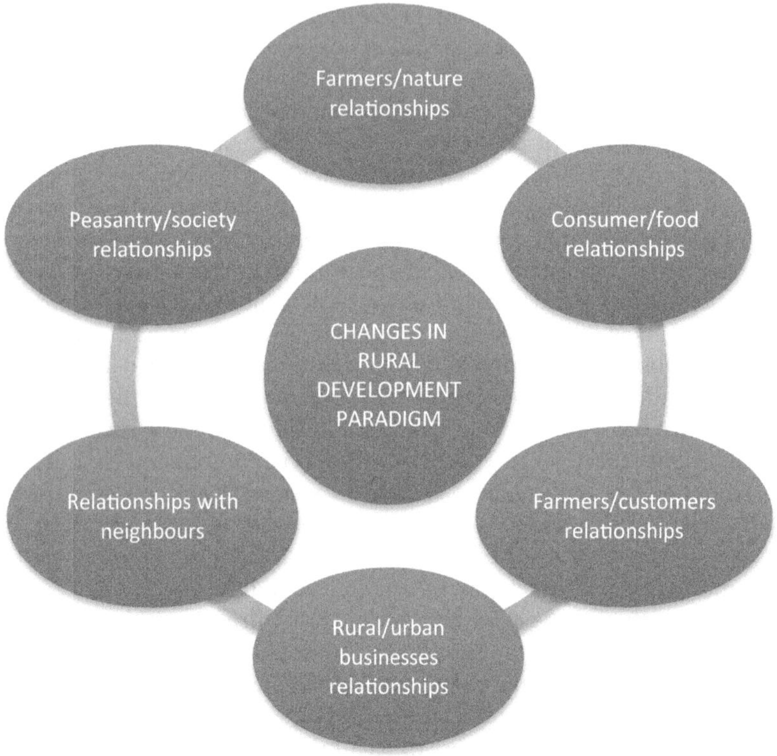

Fig. 3.2 Key aspects of change in post-industrial rural development paradigm by the prism of relationships supported by global social movements (*Source* Created by author)

3.3.1 Farmers/Nature Relationships

Industrialization converted the land into a commodity and transformed it into a guarantee for credit operations. At the same time, the industrialization of agriculture devalued and destroyed local and traditional knowledge, leading to the loss of climatically and culturally optimized production methods and seeds and to inhumane working conditions in the agricultural industry. Adherents of industrial production mode have spent the last half-century trying to eradicate or assimilate all other forms of agriculture into energy-intensive, water-guzzling, emissions-generating production model which destroys soil, habitat and species (Holt-Giménez, 2019). Agroindustrialization also had seduced farmers into extreme specializations with consequences for the environment.

In the 1970s, the organic farming movement emerged as a critique against the perceived environmentally destructive and health-threatening effects of industrial farming. Ecocentric scholars and teachers have argued that non-anthropocentrism is necessary to counter the negative impacts to environmental components that do not directly contribute to human welfare (e.g. Hourdequin, 2018; Quinn et al., 2016). But non-anthropocentric motivation is not enough. The ways to reorganize food production system should be created. According to Van der Ploeg et al., (2010), the main question is whether agricultural production is to be understood as (1) a commodity system or (2) a specific form of social and economic practices. The experience of social movements shows that within the second dimension the relationship between the local ecosystem and agricultural practice can take a variety of forms. The ideas on how to create a nature-friendly and place-relevant agricultural system is known as agroecology, an approach inspired by scientific research and indigenous practices showing how to work with the land's ecology. The sort of farming fitting under the agroecology umbrella demonstrates the possibilities to create a place-relevant agricultural ecosystem meeting the long-term rural development objectives.

The beginning of the twenty-first century represents the turning point for a new reflection on how the land needs to be reconsidered and new ways in which the neorural farmers are reconstituting themselves (Van der Ploeg et al., 2010). They argue that a farm shouldn't be a factor and food production and environmental protection must be treated as equal parts of agriculture's grand challenge Many social movements have helped farmers transition to agroecological practices (Fadaee, 2019; Meek, 2014; Raynolds, 2000). Some farmers have successfully transformed their farm while others have only put to practice some of the agroecology principles, and yet have seen a positive change. Due to their constructive experiences with agroecology many farmers have become advocates of this approach in their own area and created new social movements. For example, a growing international movement is World Wide Opportunities on Organic Farms network which offers an original form of promotion of organic farming by linking organic farmers and growers with volunteers as a way to spread cultural and educational experiences based on trust and non-monetary exchange (Mostafanezhad, 2016; Terry, 2014; Yamamoto & Engelsted, 2014).

Agroecology movements cover many farming styles today; the major global trends in creating symbiotic and synergetic effect generating relationships between farmers and nature are demonstrated by activities of such social movements and associated organizations as organic farming, permaculture, agroforestry, natural, regenerative and biodynamic agriculture, etc. They provide the platforms for conducting practice-oriented workshops for farmers and those interested in alternative agriculture, initiating pilot projects and establishing demonstration farms.

However, an agroecological transformation is curtailed by the continued dependence on corporate value chains. They have a utilitarian commitment centred on anthropocentric long-term benefits. Most consumers' behaviour patterns also were based on this worldview by focusing on cheap food with no interest on real costs. It became clear that consumer education is needed as a myopic view of highly productive agriculture helps to mask many environmental and health care costs. In the face of the success of the agroecological social movements, emerged another aspect of socio-economic life that requires transition to more synergetic relationships between consumer and food as agricultural product.

3.3.2 Consumer/Food Relationship

In the late twentieth century a significantly larger percentage of population is concerned about food safety and quality issues. The increasing citizen awareness over environmental degradation and its effects on food production has led to a shift from concerns about getting good food price to how the food was produced. As Ehrenfeld (2000, p. 204) observes: 'The challenge to industrial societies is not simply to reduce consumption, but to transform the nature of what we consume so that both human beings and natural systems can prosper'. In affluent nations, the emphasis was shifting from cheapness and quantity to quality, rarity and esteem for artisan production methods (Fernández-Armesto, 2002). This new trend is called 'quality turn' in agri-food studies (Goodman, 2003).

The desire to encourage more people to change their relationship with food became an incentive to participate in social movements. Consumer's movements strive to change markets when those markets produce value outcomes that conflict with consumers' higher order values. Most of the popular ideas of food consumers' movements originate in countries with strong gastronomic traditions and are fuelled by mounting concern that modern state and corporate institutions are unable to guarantee the socially and environmentally sound production of consumer goods. From the Nutritional Regime perspective, the industrial agri-food system is based on two dimensions: (i) it produces unhealthy food for masses through the global commodity chains; (ii) it gives quality-certified healthy food for the elites (Dixon, 2009). The dissatisfaction with both the mentioned features of the matured industrial agri-food system encouraged worldwide growth of food movements that challenge the hegemony and legitimacy of the corporate food regime. The topicality of the 'quality turn' arises together with the ageing of the society. Older people typically eat less food than younger ones, but take care about the food quality and are the most promising stakeholder group of food consumers' movements.

The actors of the food consumers' movements are not limiting themselves to the organization of protest campaigns and riots or expressions of dissatisfaction with government food policies. They have also become actively engaged in finding their own solutions how to reorganize mainstream eating lifestyle and are suggesting new eating culture which is gaining popularity in many countries. Major socio-economic innovations dealing with a healthy eating culture based on the ideas of food 'quality turn' consolidates global Slow Food mMovement. The Slow Food initiative was started in 1989 in Italy with the initial aim to defend regional traditions, good food, gastronomic pleasure and a slow pace of life. The movement approach is based on a concept of food that is defined by three interconnected principles: good, clean and fair. The stakeholders of the Slow Food movement believe that our relationship with food is tied to many other aspects of life, including culture, politics, agriculture and the environment. In over three decades of history, the movement has evolved to embrace a comprehensive approach to food that recognizes the strong connections between plate, planet, people, politics and culture. Today Slow Food represents a global social movement involving thousands of projects and millions of people in over 160 countries.

In parallel functions a variety of autonomously acting local, regional, national and international social movements focused on special aspects of food quality and eating habits, e.g. Food Security movement, Food Trust movement, Row Food movement, etc. They enable people to practice and spread in society their values (teaching children healthy eating habits, supporting farmers, reducing food waste, etc.)

3.3.3 *Farming Business/Customers Relationships*

Food nutritional quality and safety was a key for the activation of the next step in changing relationships of customers with agri-food system. The customers understood that through food choices they can collectively influence how food is cultivated, produced and distributed, and change the world as a result. The scholars point out that direct contact between farmers and consumers is an important part of a quality-oriented food culture and enables both sides to boost special qualities like traditional agricultural products, organic food, denomination of origin, etc. (Adams, 2016; Kerton & Sinclair, 2010; Starr, 2010). However, the industrialization of agriculture causes separation between areas of food production and food consumption. In many cases, the farmer and food consumer have no direct contacts.

The restoration of direct relationships between farmer and customers is in the interest of both parties. On the customer side, the restoration of direct relationships is needed for several reasons. In large urban areas, consumers often have little choice other than to initiate the connection to farmers themselves. Some do it individually, but a more easy way is to become a member of a consumer group or cooperative or other forms of buying clubs who band together to establish direct relations with farmers. Such groups in many cases succeed buy agricultural products at wholesale prices. Another important reason for direct relationships is a wish to build trust for quality assurance. The lack of trust for market provisioners—especially their perception of the deterioration of quality—combines with an endemic distrust in the state's capacity to regulate market provisioners effectively (Kjærnes et al., 2007). The quality turn has revealed the dissatisfaction with the 'impersonal "industrial food world" and a concerted turn to the "interpersonal food world" where quality conventions embed trust and tradition within a moral economy of place and provenance' (Constance et al., 2014, p. 2). The empirical studies show that consumers want to know where their food comes from and how it was produced (e.g. Bond et al., 2008; Carpio & Isengildina-Massa, 2009; Megicks et al., 2012; Morgan et al., 2006; Roman et al., 2017; Zepeda & Nie, 2012) and pay more attention not only to healthy diets, but to the impact on the environment, and local economies (Renko et al., 2014; Stanton et al., 2012; Verain et al., 2012). These concerns lead people to spend more time learning about their food sources, which motivates them to consider local foods.

On the farmer side, restoration of direct relationships is needed as a way to create new marketing channels for greater net returns, because the gap between farm gate returns and supermarket prices is often huge. Moreover, dealing directly with the producer may be seen as one way of restoring confidence in food products for some consumers (Guthrie et al., 2006). Many farmers have been able to thrive by meeting the growing demand for local food among households, restaurants, schools, health care facilities, food retailers, etc. However, it is necessary to be realistic about the ability of farmers' to alter the industrial food system by individual efforts. The creation of local food system is a complex social process. The challenge is to create a new ecosystem of local products which covers a broad network of circulation channels between farmers and food consumers and integrates product circulation during production, distribution, use and disposal processes. The creation of local products ecosystem requires innovation in marketing, management and social fields. In countering the political-economic power of the conventional industrial system, small farmers need to create a space to promote initiatives based on local food cultures and food democracy practices (Hassanein, 2003).

Empirical research has demonstrated that social movements are actively engaged in experimentation with alternatives to long food supply chains and can spur the development of local food markets (Adams, 2016; Pleyers, 2017; Starr, 2010). As a result of collaboration between consumers and producers, local food markets are often organized by consumers themselves. The consumers find a place and do logistics, while the farmers care about production and delivery services. Although the benefits of consumer groups are obvious, there are some disadvantages as well. The management and running of the group or cooperative requires a high degree of management, there are often legal matters to be considered, the group may need a formal organization with rules and regulations, and there may be staff/personnel issues to consider. The social movements help to manage the activities. They coordinate local initiatives and integrate the groups into a common network with more opportunities to access desirable food sources for consumers or find new customers for farmers. Local food movements aim to connect food producers and consumers in the same geographic region, to develop more self-reliant and resilient food networks; improve local economies; or to affect the health, environment, community or society of a particular place (Feenstra, 2002). They are the main creators of innovative ideas and practices in farming business/customers relationships building.

Many of local food movements are organized as two-sided or multi-sided platform-based networks and serves for the creation of local food markets as alternatives to global food supply systems. They provide platforms for local food including all possible types of farmer-to-consumer direct marketing, analysed by scholars (Kneafsey et al., 2013). They use for local food market development such forms as pick-your-own farms, roadside stands, farm shops, farm-based hospitality (e.g. table d'hôte, B&B), roadside sales, home delivery services, e-commerce, weekly farmers' markets, self-serve selling, farmer-owned retail outlet, food festivals/tourism events, sales directly to consumer cooperatives/ buying groups. Social movements also support and actively develop such alternative to industrial food provisioning as community-supported agriculture (Volz et al., 2016) which offers the possibility of a broad support group of people who genuinely care about the farm's survival and who are willing to share the farmer's risks. Consumers have the opportunity to connect with the earth, know and trust the people who grow their food and support the local economy. 'Farming with a face on it' was one of the first new ideas of the movement, started in Germany and Switzerland and also in Japan in the early 1960s and becoming Community-Supported Agriculture in the United States. As point out by Starr (2010, p. 482), 'this new language crystallized the concept that consumers personal relationships with their farmers could accomplish the interwoven goals of creating a more ecological society, increasing support for farmers as social actors, and verifying farming practices'. Each cultural group that adopts the Community-Supported Agriculture model shapes it differently to suit its own historical circumstances and to fit the farmers' beliefs, their land, their customers and markets.

The success of mentioned alternatives to industrial food provisioning systems forces the agri-food system towards the servitization of farming. In parallel with short food supply chain building efforts, grows a number of initiatives on service-driven business model implementation by the production of ordered amount of healthy and quality foods, instead of commodity foods supplied for the anonymous food market. Rather than focusing on a traditional transactional approach to a transfer of a product on purchase/sale basis, service-driven business model places the emphasis on optimizing value to the customer through additional add-on services.

It means that a more adaptive and responsive manner developing a relationship with the customer should be taken. The shift to a service-driven business model is especially useful for farmers involved in less known types of ecological farming such as biodynamic agriculture, natural agriculture or Fukuoka, synergistic agriculture, messianic agriculture, permaculture, etc.

3.3.4 *Rural/Urban Businesses Relationships*

After the Industrial Revolution, the balance of urban–rural relationships began to shift towards an increasing dependency of rural areas on urban economies (Davoudi & Stead, 2002). The post-industrial social movements for rural development are challenging longstanding urban–rural people business dichotomy by promoting an integrated conception of cities and countryside based on their functional interdependencies. The revision of industrial mode of rural/urban businesses relationships and experimentation with new ways is mainly focused on changing nature and role of farming as an occupation, livelihood and social activity.

Throughout history, farming was an occupation and livelihood for rural inhabitants. However, during the last several decades migration flows from the cities into rural areas are noticeable. Migrants from cities to rural areas who attempt to achieve a predominantly agrarian lifestyle have been named back-to-the-landers (Belasco, 2005). The movement 'back-to-the-land' usually associates with a social phenomenon of the 1960s and 1970s when migration from cities to rural areas took place in the United States and other countries of Global North. The pioneering phase was characterized by a multiplicity of regional-level and often unconnected initiatives (Brunori et al., 2013; Fonte & Cuccho, 2015). However, the last studies on this phenomenon points out that the idea is relevant today again and the movement 'Back-to-the-land' became an important player in the game of producing rurality/ruralities for the new millennium (Halfacree, 2007, p. 5). The ideas of the pioneers seem to have become reinvigorated in recent years, but often very different in form from that which occurred. Now the back-to-the-land movement involves people who not only are oriented in agroecological methods but also have the vision of autonomy from the conventional agri-food system. In academic literature they are named as neofarmers (Mailfert, 2007), neopeasants (Brunori et al., 2013; Van der Ploeg et al., 2010), new agrarians (Tregear et al., 2007) or new generation farmers (Milone &Ventura,

2019; Vidickiene, 2017). Most of them promote independence, interpersonal relationships, and use service-driven business model. Servitization of farming helps them to broaden the area of activities according their skills, knowledge, avocation and lifestyle. Most of the new generation farmers have come from non-agricultural lifestyles or education, and they do not seek to adopt agriculture as a full-time vocation. Gaining popularity servitization of farming by implementation of business model 'product plus service' gives an opportunity to integrate their previous skills and knowledge into farming by developing gastronomy, ecological and transformative tourism, healing, amusement and sport services closely interconnected with their agricultural production (Vidickiene et al., 2019).

In parallel to flows of new generation farmers from cities to rural areas for permanent residence in the countryside, grows initiatives on urban gardening and farming. Urban farming movement affects a paradigm shift by promoting the new vision in the rural/urban business relationships and role of farming as a social activity. The movement, born in response to a range of real needs, has become a global phenomenon, and has taken on an organized form in a large number of cities (Nicolin, 2017). The core reason for the practice of urban gardening in the Global North is the ability to bring together communities through allotments, communal gardens and other such spaces (Gorgolewski et al., 2011; Wiskereke & Viljoen, 2012). The term 'urban farming' refers to the emergence in many cities of areas cultivated by farmers who distribute the fruits of the land they work in the environs of the zone of production. In the Global South, urban farming enables citizens in deprived areas to survive by providing the urban poor with much-needed access to fresh produce (Hardman et al., 2018).

The global movement named 'Urban Farming Global Food Chain®' tries to unite local initiatives and movements of urban farming by adding one more aim which is changing industrial approaches to food security and financial security. This global movement is a part of a proactive, global campaign to Create An Abundance of Food For All In Our Generation™, along with the unique Urban Farming Coexistence Model™ that empowers people who are unemployed, underemployed, laid off, malnourished, have unhealthy diets, suffer from hunger or food insecurity. The Urban Farming Community Gardens™ which were installed during phase one of the global movement building, are called Urban Farming Food Empowerment Zones™ and the food from these gardens is free for those in need. The movement also encourages people to plant food at home and educates the community. It aims to involve into urban farming 100 million families and register their gardens as a part of the 'Urban Farming Global Food Chain®'.

3.3.5 Relations with Neighbours

The conventional view of rural areas as equivalent to agriculture is no longer reflective of the reality. In parallel with rural/urban businesses relationships, the social movements are challenging physical and social infrastructure of the industrialized countryside and are looking for alternative ways to organize rural community life. Some special radical innovations dealing with social and financial infrastructure of rural settlements are promoted by social movements focused on one aspect of the community life. The most famous and widespread are co-housing and other intentional communities movements and movements supporting various local employment and trading systems, including Timebank movement and Local Currency movements.

Most of the alternative lifestyle experiments come up with ecovillage movement which integrates all possible improvements of current lifestyle. Ecovillages are highly heterogeneous and it is impossible to describe one model that covers all cases (Dawson, 2015). If we analyse the motivations and values of ecovillage founders, we find that they have three dimensions: ecological, social and spiritual (Vidickiene, 2013). The vision for the ecological dimension of an ecovillage describes the main eco-techno values of the founders regarding buildings, infrastructure, space planning and restricted activities on the territory of the ecovillage. The wish to live in a healthy place in harmony with nature is common to the founders of all ecovillages. The vision for the social dimension of an ecovillage describes the desirable level of communality. Some of the ecovillages focus on eco-techno decisions only, with inhabitants not pursuing a high level of intercommunication: good neighbourhood relations and some social events or meetings to discuss the development of the ecovillage are sufficient. Other ecovillages aim to live as one big family; the inhabitants share land, buildings and other common resources, provide mutual aid and have a lot of common activities. The vision for the spiritual dimension of an ecovillage describes the main spiritual values of the founders. Rituals promoting the development of our inner self for harmony with the Earth and all living beings and a culture of creativity become an important tool to strengthen the spiritual life of an ecovillage.

The ideal ecovillage does not exist. However, thousands of partially successful solutions do exist. Ecovillages now can be found on each of the continents, ranging from tiny villages up to the metropolitan, the so-called inner-city ecovillages (Farkas, 2017). Local initiatives have resulted in the creation of many different types of settlements and different forms of social organization. While successful in their own regions, ecovillages are now being called to come together to play a larger role in the great transformation occurring on our planet by the Global Ecovillage Network. This international organization stands ready to deliver how ecovillages innovate, empower, educate and advocate for a regenerative world.

3.3.6 *Peasantry/Society Relationships*

All previous discussed key innovations in the relationships dealing with new paradigm of rural development aims to consolidate the concept of food sovereignty focused on the reorganization of the relationships between peasantry and society. This concept arises as an alternative to the Import Substitution Industrialization model, which has been implemented in developing countries for at least four decades. The food sovereignty idea began to be discussed in the early nineties when a new neoliberal economic policy model was implemented in many countries worldwide. In this model, state subsidies disappeared and the free market became a new development guideline.

Small farmer organizations and civil society organizations, in response to the new policies, proposed the food sovereignty approach as an alternative for the survival of agriculture in southern countries (Pachón-Ariza, 2013). Initially, the social movements in developing countries were holding actions against unfair trade practices for peasants, i.e. operated in the frame of exploitative and competitive relations and sought for the value redistribution goal. By efforts of the rural social movements in developing countries the global system of fair trade was created based on certification and several recognized fair trade certifiers appeared. Fair Trade movement does achieve many of its intended goals, although on a comparatively modest scale relative to the size of national economies. Some research indicates that the implementation of certain fair trade standards can cause greater inequalities in some markets where these rigid rules are inappropriate for the specific market (e.g. Booth & Whetstone, 2007; Carimentrand & Ballet, 2010). Other authors point out that producer benefits in many cases were close to zero because there was an oversupply of certification, and only a fraction of produce classified as fair trade was actually sold on fair trade markets, just enough to recoup the costs of certification (Dragusanu et al., 2014).

The growing criticisms about the effects of the fair trade systems led to the search for new solutions more relevant to post-industrial reality. Aiming to create more collaborative and generating synergetic effects relationships between peasantry and society, people and organizations began develope the concept of food sovereignty. Food sovereignty is a radical alternative to conventional food and agriculture development (Pimbert, 2019). 'Food sovereignty implies new social relations free of oppression and inequality between men and women, peoples, racial groups, social classes and generations' (Via Campesina, 2007). The basic premise of food sovereignty concept is that food production that is socially equitable, ecologically sustainable and adapted to the local culture, climate and soil conditions should take precedence over the export promotion for corporate profits (Weinzierl, 2019).

The reorganization of the industrial relationships between the peasantry and globalized society is crucial for successful rural development in post-industrial society. A leading role in supporting and developing the concept plays the worldwide movement 'Via Campesina' (International Peasant's Movement). The movement leaders emphasize that the food sovereignty approach contrasted against the food security approach. The concept of food security is based on the logic of the industrialization of agriculture; it aroused during the post-war period, when the paradigm of agricultural development based on food self-sufficiency in all countries was the most relevant one. The principal contention of food sovereignty is that people should be able to have more control over their own food and agriculture than the current global food system allows (Walsh-Dilley et al., 2016). Food sovereignty proponents seek fundamental social change, a transformation of society as a whole that can be achieved through the vehicle of food and agriculture. Food sovereignty constructs a proposal based on the rights of rural inhabitants beyond their production and puts food in a different context, not as a commodity. 'The consumers, beyond a fair pay for food, must offer the place the peasantry deserves in the society. When this happens, the consumers will have the right to decide the kind of food that best suits according to their viewpoints, avoiding imposed food' (Pachón et al., 2016, p. 276).

The framework for food sovereignty is evolving continually. Already three forms of sovereignty emerged that grassroots social movements mobilize around: food, energy and technological sovereignty (Altieri & Toledo, 2011). The post-industrial paradigm calls for a new relationship between peasantry and society as the so-called Grand Challenges of the twenty-first century, including global security, climate change, resource scarcity and demographic ageing are important for all members of the society and can be solved only with joint efforts of peasantry and other rural and urban population. Further scaling up and worldwide spread of sovereignty practices now depend on a capacity of social movements for rural development to apply the holistic and evolutionary approach that combines all six discussed aspects of change in relationships (Fig. 3.2) moving from competitive or exploitative relationships to collaborative and synergetic effect generating relationships.

Table 3.1 provides a summary of the discussed general trends in activities of the global social movements for rural development focused on creation and diffusion of non-technological innovations that are pushing the socio-economic system into servitization and collaborative and synergetic effect generating relationships.

The theorization of paradigm innovations-based framework offers a new and holistic way of understanding and contextualizing how and where post-industrial social movements mobilize for change. It also provides a new tool for understanding systemic creation and implementation of non-technological innovations and the role of social movements in such processes. The suggested holistic and evolutionary approach to the essence of post-industrial rural development based on three paradigm innovations leads to the following conclusions:

- Post-industrial social movements seek to encourage transition from industrial to post-industrial society. Post-industrial social movements for rural development, responding to the social, economic and environmental crises unleashed by the corporate agri-food regime, are important forces for social change based on paradigm innovations.
- The mission of post-industrial social movements for rural development is not only organization of collective protests claiming to redistribute profits, land and rights, but also practical actions for design, delivering and sustaining paradigm innovations by place-based solutions.

Table 3.1 General trends in activities of the global social movements for rural development

Major area of interest	*Non-technological innovations*	*Global movements*
Farmers/nature relationships	Agroecology	Organic, biodynamic agriculture, agroforestry, permaculture movements
Consumer/food relationships	Healthy eating culture	Slow Food movements Consumers movements
Farmers/customers relationships	Food supply chain shortening by direct marketing, including pick your-own farms, roadside stands, home delivery services, e-commerce, weekly farmers' markets, self-serve selling, and etc. Community-supported agriculture Servitization of farming	Local food movements Community-supported agriculture movements
Rural/urban businesses relationships	Lifestyle farming Servitization of rural businesses Urban gardens in Global North Urban farms in Global South	Back to the land movements Urban gardening and farming movements
Relationships with neighbours	Ecovillages, co-housing and other intentional communities Various local employment and trading systems	Ecovillage movements Co-housing movements Timebank Movements Local currency movements
Peasantry/society relationships	New role of the peasantry in the globalized society	Via Campesina movement

- The fundamental structures of post-industrial social movements are based on multiside networks that provide the platforms for reorganization of industrial exploitative or competitive relationships to post-industrial collaborative and synergetic effect generating relationships.
- After mastering networking within movements the local social movements seek to create national and international movement networks that provide the platforms scaling up paradigm innovations from a highly localized context to global level.
- The concept of paradigm innovation broadly is related to post-industrial movements and may be an umbrella and fruitful direction

for future research on all kinds of contemporary social movements, by explaining their differences in character than movements of the past, including goals, stakeholders, strategy and organizational structure.

References

Adams, R. T. (2016). Local and organic food movements. In H. Kopnina & E. Shoreman-Ouimet (Eds.), *Routledge handbook of environmental anthropology* (pp. 329–343). Routledge.

Altieri, M. A., & Toledo, V. M. (2011). The agroecological revolution in Latin America: Rescuing nature, ensuring food sovereignty and empowering peasants. *Journal of Peasant Studies, 38*(3), 587–612.

Arabska, E., Shopova, I., & Dimitrova, V. (2014). Living labs in integrated agriculture and tourism activities: Driving innovation for sustainable rural development. *Zeszyty Naukowe Małopolskiej Wyższej Szkoły Ekonomicznej W Tarnowie, 25*(2), 27–35.

Araghi, F. A. (1995). Global depeasantization, 1945–1990. *The Sociological Quarterly, 36*(2), 337–368.

Baines, T., Bigdeli, A. Z., Bustinza, O. F., Shi, V. G., Baldwin, J., & Ridgway, K. (2017). Servitization: Revisiting the state-of-the-art and research priorities. *International Journal of Operations & Production Management, 37*(2), 256–278.

Baldwin, C., & Von Hippel, E. (2011). Modeling a paradigm shift: From producer innovation to user and open collaborative innovation. *Organization Science, 22*(6), 1399–1417.

Barrett, C., Reardon, T., Swinnen, J., & Zilberman, D. (2019). *Structural transformation and economic development: Insights from the agri-food value chain revolution*. Mimeo: Cornell University.

Belasco, W. (2005). Food and the cunterculture: A story of bread and politics. In J. Watson & M. Caldwell (Eds.), *The cultural politics of food and eating: A reader* (pp. 217–234). Malden, MA and Oxford: Blackwell.

Bellandi, M., & Santini, E. (2019). Territorial servitization and new local productive configurations: The case of the textile industrial district of Prato. *Regional Studies, 53*(3), 356–365.

Bessant, J., & Tidd, J. (2007). *Innovation and entrepreneurship*. Wiley.

Blok, V., & Lemmens, P. (2015). The emerging concept of responsible innovation. In B. I. Koops (Ed.), *Responsible innovation 2* (pp. 19–35). Dordrecht: Springer.

Bond, C., Thilmany, D., & Bond, J. K. (2008). Understanding consumer interest in product and process-based attributes for fresh produce. *Agribusiness, 24,* 231–252.

Booth, P., & Whetstone, L. (2007). Half a cheer for fair trade. *Economic Affairs, 27*(2), 29–36.

Borzaga, C., & Bodini, R. (2014). What to make of social innovation? Towards a framework for policy development. *Social Policy and Society, 13*(3), 411–421.

Bosworth, G., Rizzo, F., Marquardt, D., Strijker, D., Haartsen, T., & Aagaard Thuesen, A. (2016). Identifying social innovations in European local rural development initiatives. *Innovation: The European Journal of Social Science Research, 29*(4), 442–461.

Boudreau, K. J., & Hagiu, A. (2009). Platform rules: Multi-sided platforms as regulators. *Platforms, Markets and Innovation, 1,* 163–191.

Brunori, G., Malandrin, V., & Rossi, A. (2013). Trade-off or convergence? The role of food security in the evolution of food discourse in Italy. *Journal of Rural Studies, 29,* 19–29.

Carimentrand, A. & Ballet, J. (2010). When Fair Trade increases unfairness: The case of quinoa from Bolivia (Working Papers 52010). Fonds pour la Recherche en Ethique Economique.

Carpio, C. E., & Isengildina-Massa, O. (2009). Consumer willingness to pay for locally grown products: The case of South Carolina. *Agribusiness, 25*(3), 412–426.

Chang-seok, S., Il, I., & Ki-chan, K. (2020). *Platform strategy: A new paradigm for a changing world.* World Scientific Books.

Chen, J., Yin, X., & Mei, L. (2018). Holistic innovation: An emerging innovation paradigm. *International Journal of Innovation Studies, 2*(1), 1–13.

Clunies-Ross, T., & Hildyard, N. (2013). *The politics of industrial agriculture.* Routledge.

Constance, D. H., Friedland, W. H., Renard, M. C., & Rivera-Ferre, M. G. (2014). *The discourse on alternative agrifood movements.* In M. G. Rivera-Ferre (Ed.), *Alternative agrifood movements: Patterns of convergence and divergence.* Research in rural sociology and development (Vol. 21). Emerald Group Publishing Limited.

Daly, M., & Rake, K. (2003). *Gender and the welfare state: Care, work and welfare in Europe and the USA.* Wiley.

Davoudi, S., & Stead, D. (2002). Urban-rural relationships: An introduction and brief history. *Built Environment, 28*(4), 269–277.

Dawson, J. (2015). *Ecovillages: New frontiers for sustainability.* Green Books.

Dixon, J. (2009). From the imperial to the empty calorie: How nutrition relations underpin food regime transitions. *Agriculture and Human Values, 26*(4), 321–333.

Dolata, U., & Schrape, J. F. (2016). Masses, crowds, communities, movements: Collective action in the internet age. *Social Movement Studies, 15*(1), 1–18.
Dragusanu, R., Giovannucci, D., & Nunn, N. (2014). The economics of fair trade. *Journal of Economic Perspectives, 28*(3), 217–236.
Edwards-Schachter, M., & Wallace, M. L. (2017). 'Shaken, but not stirred': Sixty years of defining social innovation. *Technological Forecasting and Social Change, 119*, 64–79.
Ehrenfeld, J. R. (2000). Sustainability and enterprise: An inside view of the corporation. In *Extended producer responsibility: A materials policy for the 21st century* (pp. 197–271). New York: Inform.
Fadaee, S. (2019). The permaculture movement in India: A social movement with Southern characteristics. *Social Movement Studies, 18*(6), 720–734.
Farkas, J. (2017). Very little heroes' history and roots of the eco-village movement. *Acta Ethnographica Hungarica, 62*(1), 69–87.
Federico, G. (2005). *Feeding the world: An economic history of agriculture, 1800-2000*. Princeton University Press.
Feenstra, G. (2002). Creating space for sustainable food systems: Lessons from the field. *Agriculture and Human Values, 19*(2), 99–106.
Fernández-Armesto, F. (2002). *Near a thousand tables: A history of food*. Simon and Schuster.
Fonte, M., & Cucco, I. (2015). The political economy of alternative agriculture in Italy. In A. Bonanno & L. Busch (Eds.), *Handbook of the international political economy of agriculture and food* (pp. 264–294). Cheltenham: Edward Elgar Publishing.
Fournier, S., & Lee, L. (2009). Getting brand communities right. *Harvard Business Review, 87*(4), 105–111.
Frouws, J. (1998). The contested redefinition of the countryside. An analysis of rural discourses in the Netherlands. *Sociologia Ruralis, 38*(1), 54–68.
Gambardella, A., Raasch, C., & von Hippel, E. (2017). The user innovation paradigm: Impacts on markets and welfare. *Management Science, 63*(5), 1450–1468.
Gebauer, H., & Binz, C. (2019). Regional benefits of servitization processes: Evidence from the wind-to-energy industry. *Regional Studies, 53*(3), 366–375.
Gentner, D., & Stevens, A. L. (1983). *Mental models*. Hillsdale, NJ: Lawrence Erlbaum Associates.
Gittell, J. H. (2002). Relationships between service providers and their impact on customers. *Journal of Service Research, 4*(4), 299–311.
Gomes, E., Bustinza, O. F., Tarba, S., Khan, Z., & Ahammad, M. (2019). Antecedents and implications of territorial servitization. *Regional Studies, 53*(3), 410–423.

Goodman, D. (2003). The quality 'turn' and alternative food practices: Reflections and agenda. *Journal of Rural Studies, 19,* 1–7.
Goodman, D. (2004). Rural Europe redux? Reflections on alternative agro-food networks and paradigm change. *Sociologia Ruralis, 44*(1), 3–16.
Goodman, R., & Peng, I. (1996). The East Asian welfare states: Peripatetic learning, adaptivechange, and nation-building. In G. Esping-Andersen (Ed.), *Welfare states in transition* (pp. 192–224). London: Sage.
Gorgolewski, M., Komisar, J., & Nasr, J. (2011). *Carrot city: Creating places for urban agriculture.* New York: Monacelli Press.
Grunwald, A. (2017). Diverging pathways to overcoming the environmental crisis. *Journal of Cleaner Production, 197,* 1854–1862.
Guthrie, J., Guthrie, A., Lawson, R., & Cameron, A. (2006). Farmers' markets: The small business counter-revolution in food production and retailing. *British Food Journal, 108*(7), 560–573.
Guzmán, J. G., del Carpio, A. F., Colomo-Palacios, R., & de Diego, M. V. (2013). Living labs for user-driven innovation: A process reference model. *Research-Technology Management, 56*(3), 29–39.
Haas, P. M. (Ed.). (1992). *Knowledge, power, and international policy coordination.* Columbia: University of South Carolina Press.
Halfacree, K. (2007). Back-to-the-land in the twenty-first century–making connections with rurality. *Tijdschrift Voor Economische En Sociale Geografie, 98*(1), 3–8.
Hardman, M., Chipungu, L., Magidimisha, H., Larkham, P. J., Scott, A. J., & Armitage, R. P. (2018). Guerrilla gardening and green activism: Rethinking the informal urban growing movement. *Landscape and Urban Planning, 170,* 6–14.
Harris, M., & Albury, D. (2009). *The innovation imperative.* London: NESTA.
Harwood, R. R. (1990). A history of sustainable agriculture. In C. A. Edwards, R. Lal, P. Madden, R. H. Miller, & G. House (Eds.), *Sustainable agricultural systems* (No. 338.162 S964, pp. 3–19). Soil and Water Conservation Society, Ankeny, Iowa (EUA).
Hassanein, N. (2003). Practicing food democracy: A pragmatic politics of transformation. *Journal of Rural Studies, 19*(1), 77–86.
Hawken, P. (2007). *Blessed unrest: How the largest social movement in history is restoring grace, justice, and beauty to the world.* Penguin.
Hazell, P., & Wood, S. (2008). Drivers of change in global agriculture. *Philosophical Transactions of the Royal Society B: Biological Sciences, 363*(1491), 495–515.
Heffernan, W., Hendrickson, M., & Gronski R. (1999). *Consolidation in the food and agriculture system* (Research Report). Washington, DC, National Farmers Union.

Hendrickson, M., & Hefferman, W. (2002). Opening spaces through relocalisation: Locating potential resistance in the weaknesses of the global food system. *Sociologia Ruralis, 42*(4), 347–369.

Hobsbawm, E. (1992). The crisis of today's ideologies. *New Left Review, 192*, 55–64.

Hoey, B. A. (2015). *International encyclopaedia of the social & behavioral sciences* (J. Wright, Ed., 2nd ed.). Boston: Elsevier.

Holliday, I. (2000). Productivist welfare capitalism: Social policy in East Asia. *Political Studies, 48*(4), 706–723.

Holt-Giménez, E. (2019). *Campesino a Campesino: Voices from Latin America's farmer to farmer movement for sustainable agriculture*. Oakland: Food First Books.

Hourdequin, M. (2018). Varieties of non-anthropocentricism: Duty, beauty, knowledge and reality. *Environmental Values, 27*(2), 113–119.

Katonáné Kovács, J., Varga, E., & Nemes, G. (2016). Understanding the process of social innovation in rural regions. *Studies in Agricultural Economics, 118*(1), 22–29.

Kerton, S., & Sinclair, A. J. (2010). Buying local organic food: A pathway to transformative learning. *Agriculture and Human Values, 27*(4), 401–413.

Kim, J. (2014). Platform business and network strategy. *STI Policy Review, 5*(1), 57–74.

Kirwan, J., Ilbery, B., Maye, D., & Carey, J. (2013). Grassroots social innovations and food localisation: An investigation of the Local Food programme in England. *Global Environmental Change, 23*(5), 830–837.

Kjærnes, U., Harvey, M., & Warde, A. (2007). *Trust in food: A comparative and institutional analysis*. Springer.

Kneafsey, M., Venn, L., Schmutz, U., Balázs, B., Trenchard, L., Eyden-Wood, T., … & Blackett, M. (2013). *Short food supply chains and local food systems in the EU. A state of play of their socio-economic characteristics* (pp. 123–129). JRC Scientific and Policy Reports.

LaVerne, E. S. (2012). *History of rural electrification and farm wiring issues*. USDA, Agricultural Research Service.

Lewis, J. (1992). Gender and the development of welfare regimes. *Journal of European Social Policy, 2*(3), 159–173.

Ludwig, D., & Macnaghten, P. (2020). Traditional ecological knowledge in innovation governance: A framework for responsible and just innovation. *Journal of Responsible Innovation, 7*(1), 26–44.

Lundvall, B. Å. (2013). Innovation studies: A personal interpretation of the state of the art. In F. Skiba (2010). *Service users as sources for innovation: An empirical study in the German services industry*. BoD–Books on Demand.

Mailfert, K. (2007). New farmers and networks: How beginning farmers build social connections in France. *Tijdschrift Voor Economische En Sociale Geografie, 98*(1), 21–31.
Marsden, T. (2003). *The condition of rural sustainability*. Assen: Uitgeverij Van Gorcum.
Marsden, T., Murdoch, J., Lowe, P., Munton, R., & Flynn, A. (1993). *Constructing the countryside*. London: UCL Press.
Martín-Retortillo M., & Pinilla V. (2012). *Why did agricultural labour productivity not converge in Europe from 1950 to 2005?* (EHES Working Papers in Economic History, No. 25, pp. 1–40). European Historical Economics Society.
Martín-Retortillo, M., & Pinilla, V. (2013). *Patterns and causes of growth of European agricultural production, 1950-2005* (No. 1302). Asociación Española de Historia Económica.
Meek, D. (2014). Agroecology and radical grassroots movements' evolving moral economies. *Environment and Society, 5*(1), 47–65.
Megicks, P., Memery, J., & Angell, R. J. (2012). Understanding local food shopping: Unpacking the ethical dimension. *Journal of Marketing Management, 28*(3/4), 264–289.
Meynhardt, T., Chandler, J. D., & Strathoff, P. (2016). Systemic principles of value co-creation: Synergetics of value and service ecosystems. *Journal of Business Research, 69*(8), 2981–2989.
Miles, I. (2008). Patterns of innovation in service industries. *IBM Systems Journal, 47*(1), 115–128.
Milone, P., & Ventura, F. (2019). New generation farmers: Rediscovering the peasantry. *Journal of Rural Studies, 65*, 43–52.
Morgan, K., Marsden, T., & Murdoch, J. (2006). *Worlds of food place, power, and provenance in the food chain*. Oxford University Press.
Mostafanezhad, M. (2016). Organic farm volunteer tourism as social movement participation: A Polanyian political economy analysis of World Wide Opportunities on Organic Farms (WWOOF) in Hawaii. *Journal of Sustainable Tourism, 24*(1), 114–131.
Mulgan, G., Tucker, S., Ali, R., & Sanders, B. (2007). *Social innovation: What it is, why it matters, and how it can be accelerated*. Skoll Centre for Social Entrepreneurship, SAID Business.
Navarro, F., Labianca, M., Cejudo, E., de Rubertis, S., Salento, A., Maroto, J. C., & Belliggiano, A. (2018). Interpretations of innovation in rural development. The cases of leader projects in Lecce (Italy) and Granada (Spain) in 2007–2013 period. *European Countryside, 10*(1), 107–126.
Nemes, G. (2005). *Integrated rural development. The concept and its operation* (No. MT-DP-2005/6). IEHAS Discussion Papers.

Neumeier, S. (2017). Social innovation in rural development: Identifying the key factors of success. *The Geographical Journal, 183*(1), 34–46.

Nicolin, P. (2017). Urban farming movement. In *Design innovations for contemporary interiors and civic art* (pp. 186–189). IGI Global.

OECD Policy Brief. (2006). *Reinventing rural policy*. Paris: OECD.

Pachón, F., Bokelmann, W., & Ramírez, C. (2016). Rural development thinking, moving from the green revolution to food sovereignty. *Agronomía Colombiana, 34*(2), 267–276.

Pachón-Ariza, F. A. (2013). Food sovereignty and rural development: Beyond food security. *Agronomía Colombiana, 31*(3), 362–377.

Palmatier, R. W. (2008). *Relationship marketing*. Cambridge, MA: Marketing Science Institute.

Pimbert, M. (2019). *Sustainable food systems and agriculture*. In Encyclopedia of food security and sustainability.

Pleyers, G. (2017). The local food movement in Belgium: From prefigurative activism to social innovations. *Interface: A Journal for and About Social Movements*, *9*(1), 123–139.

Pol, E., & Ville, S. (2009). Social innovation: Buzz word or enduring term? *The Journal of Socio-Economics, 38*(6), 878–885.

Quinn, F., Castéra, J., & Clément, P. (2016). Teachers' conceptions of the environment: Anthropocentrism, non-anthropocentrism, anthropomorphism and the place of nature. *Environmental Education Research, 22*(6), 893–917.

Ramaswamy, V., & Ozcan, K. (2014). *The co-creation paradigm*. Stanford University Press.

Raynolds, L. T. (2000). Re-embedding global agriculture: The international organic and fair trade movements. *Agriculture and Human Values, 17*(3), 297–309.

Reardon, T., & Barrett, C. B. (2000). Agroindustrialization, globalization, and international development: An overview of issues, patterns, and determinants. *Agricultural Economics, 23*(3), 195–205.

Reardon, T., Timmer, C. P., Barrett, C. B., & Berdegué, J. (2003). The rise of supermarkets in Africa, Asia, and Latin America. *American Journal of Agricultural Economics, 85*(5), 1140–1146.

Renko, S., Lang, M., Stanton, J., & Qu, Y. (2014). Consumers' evolving definition and expectations for local foods. *British Food Journal, 116*(11), 1808–1820.

Roman, S., Sánchez-Siles, L. M., & Siegrist, M. (2017). The importance of food naturalness for consumers: Results of a systematic review. *Trends in Food Science & Technology, 67*, 44–57.

Rosin, C., Stock, P., & Campbell, H. (Eds.). (2013). *Food systems failure: The global food crisis and the future of agriculture*. Routledge.

Rozelle, S., & Swinnen, J. F. M. (2004). Success and failure of reform: Insights from the transition of agriculture. *Journal of Economic Literature, 42*(2), 404–456.

Scott-Smith, T. (2016). Humanitarian Neophilia: The innovation turn and its implications. *Third World Quarterly, 37*(12), 2229–2251.

Seyfang, G., & Haxeltine, A. (2012). Growing grassroots innovations: Exploring the role of community-based initiatives in governing sustainable energy transitions. *Environment and Planning C: Government and Policy, 30,* 381–400.

Seyfang, G., & Smith, A. (2007). Grassroots innovations for sustainable development: Towards a new research and policy agenda. *Environmental Politics, 16,* 584–603.

Smith, A., Fressoli, M., & Thomas, H. (2013). Grassroots innovation movements: Challenges and contributions. *Journal of Cleaner Production, 63,* 114–124.

Smith, A., Hargreaves, T., Hielscher, S., Martiskainen, M., & Seyfang, G. (2016). Making the most of community energies: Three perspectives on grassroots innovation. *Environment and Planning A, 48*(2), 407–432.

Song, L. Z., Song, M., & Di Benedetto, C. A. (2009). A staged service innovation model. *Decision Sciences, 40*(3), 571–599.

Stanton, L. J., Wiley, J. B., & Wirth, F. F. (2012). Who are the Locavores? *Journal of Consumer Marketing, 29*(4), 248–261.

Starr, A. (2010). Local food: A social movement? *Cultural Studies? Critical Methodologies, 10*(6), 479–490.

Suhartanto, D., Helmi Ali, M., Tan, K. H., Sjahroeddin, F., & Kusdibyo, L. (2019). Loyalty toward online food delivery service: The role of e-service quality and food quality. *Journal of Foodservice Business Research, 22*(1), 81–97.

Sumelius, J., & Bäckman, S. (2008). Review of studies on the establishment and management of policies for multifunctionality. *International Journal of Agricultural Resources, Governance and Ecology, 7*(4/5), 386–398.

Terry, W. (2014). Solving labor problems and building capacity in sustainable agriculture through volunteer tourism. *Annals of Tourism Research, 49,* 94–107.

Torre, A., & Wallet, F. (2016). *Regional development in rural areas: Analytical tools and public policies.* Springer.

Tregear, A., Arfini, F., Belletti, G., & Marescotti, A. (2007). Regional foods and rural development: The role of product qualification. *Journal of Rural Studies, 23*(1), 12–22.

Trischler, J., Johnson, M., & Kristensson, P. (2020). A service ecosystem perspective on the diffusion of sustainability-oriented user innovations. *Journal of Business Research, 116,* 552–560.

Van der Ploeg, J. D., & Long, A. (Eds.). (1994). *Born from within. Practice and perspectives of endogenous rural development*. Assen: Van Gorcum.

Van der Ploeg, J. D., & van Dijk, G. (Eds.). (1995). *Beyond modernization. The impact of endogenous rural development*. Assen: Van Gorcum.

Van der Ploeg, J. D., Ye, J., & Schneider, S. (2010). Rural development reconsidered: Building on comparative perspectives from China, Brazil and the European Union. *Rivista Di Economia Agraria, 65*(2), 163–190.

Van der Ploeg, J. D., et al. (2000). Rural development: From practices and policies towards theory. *Sociologia Ruralis, 40*(4), 391–408.

Vendrell-Herrero, F., & Wilson, J. R. (2017). Servitization for territorial competitiveness: Taxonomy and research agenda. *Competitiveness Review, 27*(1), 2–11.

Verain, M. C. D., Bartels, J., Dagevos, H., Sijtsema, S. J., Onwezen, M. C., & Antonides, G. (2012). Segments of sustainable food consumers: A literature review. *International Journal of Consumer Studies, 36*(2), 123–132.

Via Campesina. (2007). *The declaration of Nyéléni*. https://nyeleni.org/IMG/pdf/DeclNyeleni-en.pdf. 18 September 2020.

Vidickiene, D. (Ed.). (2013). *Living in harmony: Inspiring stories from ecovillages*. Vilnius: Lithuanian Institute of Agrarian Economics.

Vidickiene, D. (2017). Economic regulation directions regarding the essential shift in the structure of economy. *Public Policy and Administration, 16*(3), 468–481.

Vidickiene, D., & Gedminaite-Raudone, Z. (2018). Challenges for agricultural policy in the service driven economic system. *Economics of Agriculture, 65*(4), 1545–1555.

Vidickiene, D., & Melnikiene, R. (2014). *Evolution of rural policy*. Vilnius: Lithuanian Institute of Agrarian Economics.

Vidickiene, D., Gedminaite-Raudone, Ž., & Simonaityte, V. (2019). *Servitization of farming: Business model "product plus service" in Lithuania*. Vilnius: Lithuanian Institute of Agrarian Economics.

Volz, P., Weckenbrock, P., Nicolas, C., Jocelyn, P., & Dezsény, Z. (2016). *Overview of community supported agriculture in Europe*. European CSA Research Group.

Walsh-Dilley, M., Wolford, W., & McCarthy, J. (2016). Rights for resilience: Food sovereignty, power, and resilience in development practice. *Ecology and Society, 21*(1).

Weatherell, C., Tregear, A., & Allinson, J. (2003). In search of the concerned consumer: UK public perceptions of food, farming and buying local. *Journal of Rural Studies, 19*(2), 233–244.

Weinzierl, C. (2019). The history of the CAP in the context of the international agricultural and trade regime. In *How the EU is destroying smallholder agriculture inside and outside Europe*. Attac Austria, Vienna.

Wells, P. (2018). Degrowth and techno-business model innovation: The case of Riversimple. *Journal of Cleaner Production, 197,* 1704–1710.
Wenger, E. (1998). Communities of practice: Learning as a social system. *Systems Thinker, 9*(5), 2–3.
Wiskerke, J. S., & Viljoen, A. (2012). Sustainable urban food provisioning: Challenges for scientists, policymakers, planners and designers. In J. S. Wiskerke, & A. Viljoen (Eds.), *Sustainable food planning: Evolving theory and practice* (pp. 19–35). Wageningen Academic Publishers: Wageningen, The Netherlands.
Wolf, S. A., & Bonanno, A. (Eds.). (2013). *The neoliberal regime in the agri-food sector: Crisis, resilience, and restructuring*. Routledge.
World Bank. (2013). *Agriculture and rural development*. World Bank Data. Retrieved from http://data.worldbank.org/topic/agriculture-and-rural-development. 18 November 2020.
Yamamoto, D., & Engelsted, A. K. (2014). World Wide Opportunities on Organic Farms (WWOOF) in the United States: Locations and motivations of volunteer tourism host farms. *Journal of Sustainable Tourism, 22*(6), 964–982.
Zavratnik, V., Superina, A., & Stojmenova Duh, E. (2019). Living labs for rural areas: Contextualization of living lab frameworks, concepts and practices. *Sustainability, 11*(14), 3797.
Zepeda, L., & Nie, C. (2012). What are the odds of being an organic or local food shopper? Multivariate analysis of U.S. food shopper lifestyle segments. *Agriculture and Human Values, 29*, 1–14.

CHAPTER 4

Actors of New Rural Social Movements as Agents of Change

Vitalija Simonaitytė

4.1 Rural Population as Agents of Rural Development

Rural population includes a lot of different actors, i.e. small and large farmers, agribusiness, agricultural companies, agricultural cooperatives, large-scale retailers and many others. In the light of (new) social movements, rural population actors can be divided into two broad groups: actors representing and acting as agents of industrial movements and post-industrial movements. As it was already discussed in Chapter 3, the proponents of the post-industrial discourse among rural populations are mainly small farmers and rural entrepreneurs and to some extent agricultural enterprises and agricultural cooperatives. While large farmers and movements representing them are more likely to keep rural policies stable and make little contribution to progress regarding rural development and (social) innovations. Usually, their purpose is to bend the redistribution of public goods in their favour and to maintain the status quo. The scope of this chapter is to discuss the most important rural population agents, such as small farmers, agricultural companies and agricultural cooperatives and their goals, level of organization, ways in which organizations operate and how they change the rural areas and their development. So, the question is not only how farmers and agribusiness companies align with new rural

R. Vilkė et al., *Rural Economic Developments and Social Movements*, Palgrave Advances in Bioeconomy: Economics and Policies, https://doi.org/10.1007/978-3-030-71983-8_4

social movements but also how small farmers, agricultural enterprises and cooperatives influence and change rural development.

Small farmers. There were 10.5 million farms in the EU in 2016, with the vast majority of these (95.2%) classified as family farms. In 2016, family farms accounted for around 80% of the labour force input and around 60% of the total utilized agricultural area, of livestock units and of the value of the agricultural output (Agriculture Statistics, 2016). Small farmers are the most important actors of rural population as well as their goals are the broadest in changing rural development. Even the small farmers usually are passive players and the majority among them is unorganized and lacks political experience and channels to challenge existing status quo, their role is extremely important especially in creating and developing local (products and related services) ecosystems. One of the most important goals of farmers is satisfying the food needs of the population and supporting economic development. There is no doubt that this particular goal stays the most important one until nowadays, however, this particular goal draws an essential difference between traditional (usually large) farmers and post-industrial farmers and post-industrial rural development.

Traditional farmer's main goals for many years have been satisfying the food needs of the population, supporting economic development and providing employment in rural areas. However in the twenty-first century without denying the importance of satisfying the food needs of the population, small farmers are becoming niche farmers which more and more often are practicing sustainable agriculture and valorizing traditional products as well as implementing the valorization of products for which there is a comparative advantage and the transition from primary to agro-industrial production. In reaching these goals, associations and various movements of small farmers, family farms and cooperatives play a crucial role by representing their members, advocating for policies which are in favor of their interests, framing the political agenda, bringing public attention to certain issues, educating society and many others initiatives.

As Van der Ploeg and Roep (2003) described, European rural small agricultural enterprises' development can be seen in three directions: broadening rural area, deepening agro-food supply chain and regrounding mobilization of resources. It must be said that this approach is suitable for all agents of rural development and the case of small farmers proves that as they are active in all three of these directions and represent rural population to a great extent.

Broadening rural areas. Broadening the understanding of rural areas can follow different trajectories. It could be agrotourism, the management of nature and landscape (Baldock & Baeufoy, 1993; Renting & Van der Ploeg, 2001), to the development of new on-farm activities (e.g. care activities) and diversification (as, for example, the production of energy; Knickel & Renting, 2000) as well as servitization. Taken together, such activities enlarge the income flows, while they simultaneously imply the delivery of goods and services society is willing to pay for (as cited in Van der Ploeg & Roep 2003).

Deepening the agro-food supply chain. Small farmers and their organizations in the twenty-first century are developing alternative farm practices, farmers and eaters are engaged in community-supported agriculture, groups of farmers and consumers are working to guarantee the right to a nutritious and sufficient diet, and as policy advocates are operating at the national and international levels (Hassanein, 2003, p. 77; Henderson, 2000). It is clear that small farmers are no longer just a primary part of the chain in growing-processing-producing food but they are seeking and getting more involvement in the whole food supply chain.

Regrounding the mobilization of resources. These are pluri-activity and farming economically. Through pluri-activity (Bryden et al., 1992; Fuller & Brun, 1991) the farm enterprise is partly built on off-farm income. This implies the maintenance of a farm that would otherwise probably disappear. It also implies that the farms concerned become less dependent on the tendencies, trends and variations in the big commodity markets. In the past, pluri-activity could be considered as an expression of poverty. It is, however, currently becoming increasingly an expression of the opposite and a well-thought-out strategy. For many people, it is the preferred combination for living in the countryside, having a farm and simultaneously having an urban job and the associated income security (as cited in Van der Ploeg & Roep, 2003, p. 43).

Analysing small farmers goals in the spite of post-industrial movements, it also has to be said, that small farmers often implements other three very important effects: firstly, new forms of agriculture are changing farmer and nature relationship; secondly, new forms of accessibility to food for consumers are changing farmer and customer relationships, and thirdly farmers and customers relationship and even cooperation is changing rural and urban (businesses) relationships. How does this work from a theoretical perspective? (Fig. 4.1).

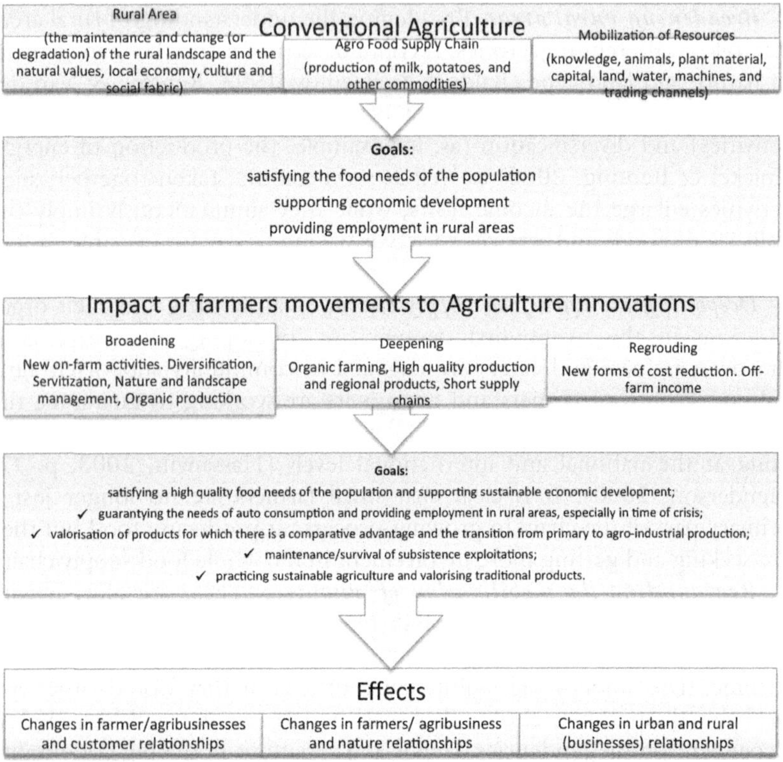

Fig. 4.1 Changing goals and effects of agriculture in industrial and post-industrial agriculture (*Source* Developed by authors based on Van der Ploeg and D. Roep [2003])

As it was described in Chapter 3, because of changing situation and a high competitiveness between small and large farmers, the latter had to find a spot to offer better or even exceptional products. By broadening rural areas, i.e. introducing agri-tourism, diversification, servitization, educational practices held by farmers, farmers and customers had to restore a direct relationship between farmer and customers, which was in the favour of both parties. It shows that the goal of broadening the products and services of rural areas, had an excellent effect on *farmers and customers relationships*.

Secondly new forms of agriculture, agroecology movements—organic, biodynamic agriculture, agroforestry, permaculture initiatives, natural, regenerative and biodynamic agriculture, etc., are changing *farmer and nature relationships*, where land, water, soil are no longer seen as a never-ending resources but as a tool to reach a sustainable coexistence by not exploiting nature. Looking at this from Van der Ploeg and Roep (2003) perspective, it is clear that their described goal of deepening the agro-food supply chain has an effect of changing farmer and nature relationship.

Finally the regrounding the mobilization resources, when small farmers have to diversify their activities and are only part-time farmers still having paid job in urban areas, connects *rural and urban (business) relationship* and have an important effect on rural development.

Analysing small farmers as actors of rural population who are influencing changes of rural development, it is important to ask what do small farmers seek and how they (can) influence decision-making or public policy overall? As a primary purpose of agricultural activities is economical—i.e. to earn for a living, the other goals of farmers activities have been discussed above, the unanswered question stays how small farmers can influence public life and decision-making. Even small farmers are not considered to have the capacity to assume political power in the absence of concerted efforts to get them involved and organized, however, they do tend to organize to various movements—local food, agroecology, organic, biodynamic, agroforestry, etc. Depending on the level of organization, local food and agroecology movements, as any organization have their goals and functions (participation, representation, education, advocacy, agenda-setting, monitoring, etc.), which usually are broader when just economic issues related to their direct occupation. These aspects of new social movements, as well the ones uniting small farmers are discussed in Chapters 6 and 7.

Agricultural enterprises and businesses. Agricultural enterprises and businesses are another important actors of rural population which are changing rural development. However, talking about agribusinesses there must be made a clear distinction between large and small agricultural businesses. As it was discussed in Chapter 3, fundamental social restructuring as industrialization gave rise to new farmers' elite representing large agribusinesses while worsening the situation of small farmers and small agribusiness as that is one of the reasons why large agribusinesses are still very conventional, intensive and usually are tended to exploit all the available recourses. Because of that, when talking about rural development and rural changes in spite of new social movements, it is clear that only small agricultural enterprises and businesses could be willing and could be able to act as agents of rural development.

Taking into account Van der Ploeg and Roep's (2003) research on agricultural enterprises three directions—broadening rural area, deepening agro-food supply chain and regrounding mobilization of resources—of rural development are employed for further analysis.

Broadening rural areas. The same as in the case of small farmers, agricultural enterprises and businesses can employ broadening of rural areas in different trajectories. Agribusinesses use agrotourism, the management of nature and landscape to the development of new on-farm activities (e.g. care activities, education, tastings) and diversification as, for example, the production of energy; as well as servitization. The servitization of rural businesses can be the most progressive way to respond to major rural development challenges, especially for generating desirable changes in industrial food supply chains because service-driven farming business model reduces food demand and waste and provokes desirable behavioural changes of consumers. Taken together, such activities as diversification and servitization, enlarge the income flows, while they simultaneously imply the delivery of goods and services society is willing to pay for. So, the differences between small farmers and small agribusinesses are minimal in spite of the activities broadening rural areas.

Deepening the agro-food supply chain. Small agribusinesses usually are very similar to small farmers, as they are developing alternative farm practices, farmers and eaters are engaging in community-supported agriculture, groups of agricultural enterprises and consumers are working to guarantee the right to nutritious and sufficient food (Hassanein, 2003). It is clear that small agribusinesses, as well as small farmers, are no longer just a primary part of the chain in growing-processing-producing food but they are seeking and getting more involvement in the whole food supply chain.

Regrounding the mobilization of resources. There can many different resources needed in small agribusiness, where the most important are natural resources, as well as labour, time, knowledge, skills, community, cooperation and many others. As many of these resources are limited, farmers of small agricultural businesses usually have to be very creative and employ new forms of work organization. For example, some of the small agribusinesses are outsourcing many activities, e.g. soil tillage, sowing, etc., as there is no need to find suitable employees and provide them with a full-time job on a daily basis, and the quality of services provided by companies is higher than when hiring the rural population an finally it helps them to farm more efficiently (Vidickienė et al., 2019, p. 46).

Finally, there is a question how these three strategies of broadening, deepening and regrounding strategies implemented by agribusinesses are affecting rural areas and their development in spite of new social movements and what are their effects changing paradigm of rural development?

By broadening rural areas, i.e. introducing agri-tourism, diversification, servitization, educational practices held by small agribusinesses and customers had to restore a direct relationship between *agribusinesses and customers*, which was in the favour of both parties. The servitization can be an important tool for changes in rural development as service-driven business models consider how value may be influenced and realized via more dynamic, interactive arrangements between consumers and farmers and it shows that the goal of broadening the products and services of rural areas, have an excellent effect on small agribusinesses and customers relationships.

By deepening the agro-food supply chain, many new forms of agriculture, agroecology movements—organic, biodynamic agriculture, agroforestry, permaculture initiatives, natural, regenerative and biodynamic agriculture—emerged which are changing *small agribusiness and nature relationships*, where land, water, soil are no longer seen as a never-ending resources but as a tool to reach a sustainable coexistence by not exploiting nature. The deepening of the agro-food supply chain provides the platforms for conducting practice-oriented workshops for agribusinesses and consumers who are interested in alternative agriculture, initiating pilot projects or establishing demonstration farms it creates and empowers the better and deeper relationships between small agribusinesses and nature.

Finally, the regrounding the mobilization resources, when small farmers have to diversify their activities, employ new business model, e.g. service-driven business model when they are not only providing services but also actively buying them, connects *rural and urban business relationship* and have an important effect on rural development.

Agricultural cooperatives and local food movements. Agricultural cooperatives and local food movements are the third actor changing rural development and acting as active and influential actors. Why cooperate and what are cooperatives and local food movements? Looking from a traditional perspective, agricultural cooperatives represent a lever of rural development, i.e. one of the institutions that can be used to increase economic, cultural and social capital in an area, as well as agricultural cooperatives stimulate (self) employment. However, looking at agricultural cooperatives and other local food movements from post-industrial perspective, these actors are much more complex than ever before and their role in rural development is getting more important.

Local food movements are getting more important and more influential movements in all European rural areas as these community-supported agricultural movements helps to get a better price for both farmers and consumers and moreover it builds direct relationships and trust between farmers and consumers. Establishment of cooperatives and local food movements overcome very similar challenges from both customers and small farmers or small agribusinesses—for customers cooperation and other forms of buying initiatives helps to establish direct relationships with farmers while on the other hand cooperatives and local food movements helps farmers to reach their customers, reduces expenses in reaching for the customers, logistics costs, etc. However, what are the goals of the agricultural cooperatives and local food movements and what effect do they have on rural development?

One of the most recent analysis is made by Ortiz-Miranda et al. (2010) who analysed agricultural cooperatives in the region of Valencia employing Van der Ploeg's and Roep's (2003) model of three strategies—deepening, broadening and regrounding.

Broadening rural areas. Agricultural cooperatives as well as others agents of rural population employ broadening strategies of rural areas. According to Ortiz-Miranda et al. (2010, p. 667), cooperatives in Valencia are using at least four broadening strategies, such as—providing associates with non-agricultural services (e.g. cooperatives' facilities and staff which enlarge traditional activities of many cooperatives), providing nonassociates with non-agricultural services (i.e. providing local stores, supermarkets, supply of inputs for garden), energy production and cooperative tourism (thematic routes, visiting farms, tasting activities, etc.). Taken together, such activities enlarge the income flows, while they simultaneously imply the delivery of goods and services society is willing to pay for.

Deepening the agro-food supply chain. The main deepening strategies employed by cooperatives are food processing, territorial label based on the differentiation of raw materials, organic certification and direct selling of high-quality cooperative products. These activities allow the optimization of the work of the cooperative, deepens the cooperation and the involvement in agro-food supply chain and most importantly it adds value to food production and valorizes local exceptional quality products (Ortiz-Miranda et al., 2010, p. 667).

Regrounding the mobilization of resources. As it was mention above, reorganization of resources mean that organization or person relocate available resources, which can be natural resources, money, staff, time, knowledge, skills, etc. Case studies made by Ortiz-Miranda et al. (2010, p. 669) revealed at least three regrounding strategies employed by cooperatives—first, direct management, i.e. a reallocation between the individual farm's labour force (either family based or hired), and the cooperative's labour force; second, centralized production planning when cooperative's staff plans farmers' production (what, when and how it is produced) means a reallocation of decision-making—which is also a reallocation of transaction costs; third, utilization of animal waste in organic farming fertilization.

Finally, these strategies employed by cooperatives and local food movements can have rural implications. Broadening strategy illustrates cooperatives' aspiration in searching for economies of scope to increase the income and profit. Because a high competitiveness between small and large farmers, cooperatives had to broaden their set products and even services by introducing agri-tourism, visiting farms, tasting activities which are allowing the valorization of cooperatives' productions and which led to establish a direct relationship between agricultural *cooperatives and customers*, which is in favour of both parties.

By employing deepening strategy cooperatives try to link their new (processed, quality) products to a territorial origin, either directly (i.e. emphasizing their origin with a Protected Designation of Origin), or indirectly (i.e. stressing the local feature of autochthonous varieties). In both cases, these actions underpin the creation of a 'territorial label' that could become a valuable asset for other economic—even non-agricultural—activities. On the other hand, these strategies facilitate the creation or strengthening of economies of synergy among cooperatives and other economic agents in rural areas. These case studies show some of the possibilities that arise, such as the linkage of food quality improvement to tourism and restaurants, and the related processing of products (Ortiz-Miranda et al., 2010, p. 667). Looking at this from Van der Ploeg and Roep (2003) perspective, it is clear that their described goal of deepening the agro-food supply chain has an effect of changing *cooperatives as agribusiness and nature relationship* as well as it changes and influence *relationships between cooperatives as agribusinesses and customers.*

Regrounding the mobilization of resources revealed even more interesting aspects of effects made by cooperatives to rural development, as new regrounding practices are changing the internal structure of cooperatives' organization and decision-making process. Even at this stage of the research there is not enough evidence to state that regrounding the mobilization of resources is changing the relationship between any urban or rural actors, but it is changing the structure of cooperatives which implications on further development of rural areas cannot be recognized yet.

To sum up. What is the end goal and where are we going to? Agents of rural population, such as small farmers, small agribusinesses and enterprises, and agricultural cooperatives are important in rural development by accomplishing not just economic goals but acting in a much more wider spectrum of issues by satisfying the food needs of the population and supporting economic development; guarantying the needs of auto consumption and providing employment in rural areas, especially in time of crisis; valorization of products for which there is a comparative advantage and the transition from primary to agro-industrial production; maintenance/survival of subsistence exploitations; practicing sustainable agriculture and valorizing traditional products. It proves that integrated rural development is a new and solid paradigm, which is strongly rooted in and empowered by practice where rural population is active and not only open to new trends but is creating them. Analysis of agents of rural population revealed that one of the essential aspects of post-industrial paradigm is the creation and consolidation of new interlinkages between rural and urban actors and between agriculture and society at large. Framework for post-industrial rural development paradigm building where non-technological drivers of progress, service-driven business model and collaborative and synergic relationships coexist and empower each other are essential and agents of rural population shows their existence and their importance in changing rural areas. That is why the end goal for rural population is rural development, rural welfare empowering social, economic and cultural development and the rural development is not possible without non-technological drivers of progress, service-driven business model and especially collaborative and synergistic relationships.

4.2 Urban Population as Agents of Rural Development Paradigm Shift

Urban population, as well as earlier discussed rural population, includes a lot of different actors which are active in the development of rural areas. Even the variety of urban actors is exceptionally wide and boundaries between urban and rural agents are merging, the most important agents of rural development are groups and movements of consumers, back-to-the-land movements and various new forms of rural lifestyle, such as ecovillages or co-housing movements inspired by urban population. These actors are not only acting as agents of rural development, but more importantly these agents are enabling the rural development paradigm shift.

The scope of this chapter is to discuss the most important and influent actors of urban populations, namely consumers, back-to-the-land movements and movements changing rural lifestyle, e.g. ecovillages; and secondly, it is important to discuss the goals of above-mentioned actors, ways in which organizations operate and how they change the rural areas and their development. The most important question is not only how consumers, back-to-the-land movements and ecovillages align with new rural social movements but also how these agents of urban population influence and change rural development and more important—how do they change the rural development paradigm.

Consumers. The organized consumer's movement started and flourished in urban areas at the end of the twentieth century when more and more people became concerned about food safety and food quality issues. As rural population has a stronger sense of identity to family and community, the urban population usually is much more fragmented, diverse, manifold and distanced from its food. Because of this reason to achieve any organized goal, agents of urban population have to organize and establish various movements in order to reach their goals. That is why there is a huge variety of consumers movements—one of them is concerned about organic food, others about fair, clean, ethically grown or even socially responsible food. As Moore (2006, p. 416) quoted 'If there is an imaginary pendulum swinging from nature to industrial society, then organic food represents a move back towards nature or more natural food (Murdoch et al., 2000)'. Health, safe and tasty food are usually the main reasons of consumers to elaborate and to develop formal or informal consumers' movement. However, there could be even more complex reasons, for example, to eat organic food—Makatouni's (2002) revealed that interrelated human, animal and environment life values are also very important for organic food consumers.

Consumers movements also emphasize the shift from cheapness and quantity to quality, rarity and esteem for artisan production methods (Fernandez-Armesto, 2001). This new trend is called 'quality turn' in agri-food studies (Goodman, 2003). However, the 'quality turn' and desire for high-quality, clean, fair and/or organic food have various implications and effects on rural development. So again, by taking into account Van der Ploeg and Roep's (2003) research on three directions—broadening rural area, deepening agro-food supply chain and regrounding mobilization of resources—of rural development are employed for further analysis.

Broadening rural areas. Consumers' movements as agents of urban population employ broadening strategies of rural areas in many ways. Consumers movements together with farmers create local food markets or even online platforms to buy or exchange products. Slow Food International is one of such kind initiatives, where the lead to create a local community can come from both—consumers and farmers or agribusinesses. Conscious and responsible consumers participate in on-farm activities and there are many cases, where such cooperation is initiated by consumers and their movements. Even all these activities must be supported by farmers and overall rural population, there are many cases where urban population by suggesting new ideas and proposals to urban population, affect not only their relationships into deeper cooperation but also influence changes in rural development and overall the perception of rural areas. However, the strategy of broadening rural areas is very interrelated with deepening the agro-food chain, as many broadenings serves as marketing for short food supply chain.

Deepening the agro-food supply chain. Deepening the agro-food supply chain is the main area, where consumers have the biggest influence on the rural population and where their cooperation and their relationships overlap the most. According to Marsden, Banks and G. Bristow's (2000) definition of short food supply chain is an umbrella term which enables the consumer to confidently make connections and associations with the place/space of production, and, potentially, the values of the people involved and the production methods employed. Marsden et al. (2000, pp. 425–426) distinguish three types of short food supply chain: (1) Face-to-face: consumer purchases a product directly from the producer/processor on a face-to-face basis. Authenticity and trust are mediated through personal interaction. The Internet also now presents opportunities for a variant of face-to-face contact through online trading and web pages; (2) Spatial proximity: products are produced and retailed in the specific region (or place) of production, and consumers are made aware of the 'local' nature of the product at the point of retail; (3) Spatially extended: where value and meaning laden information about the place of production and those producing the food is translated to consumers who are outside of the region of production itself and who may have no personal experience of that region. Consumers movements focus on the relocalization of local food system towards a direct relationship between producers and consumers, where a shift from 'impersonal "industrial food world" to the "interpersonal food world" is essential and where quality conventions embed trust and tradition within a moral economy of place and provenance' (Constance & Renard, 2014, p. 2). Following a food regime perspective, 'Food from Somewhere' builds on the notion of food sovereignty and the processes of relocalization, challenging the 'Food from Nowhere' (Campbell, 2009) (as cited in Orria & Luise, 2017, p. 138). Consumers movements are very important agents in empowering short food supply chains, as personal relationship, trust, face-to-face contact and knowledge about the product are very important aspects of customers and farmers, and customers and food relationships in satisfying high-quality food needs, empowering sustainable agriculture and valorizing traditional products.

Regrounding the mobilization of resources. There can be many different resources in consumers movements, where the most important are human resources, such as labour, time, knowledge, skills, community, cooperation and many others. As many of these resources are limited, consumers movements must be innovational and employ new forms of work organization, employ online communication or even online markets. The second way of regrounding the mobilization of resources is much more marginal and usually it is protest campaigns and riots or expressions of dissatisfaction with national food policies.

Finally, there is a question how these three strategies of broadening, deepening and regrounding strategies implemented by consumers' movements are affecting rural areas and rural development paradigm shift? By broadening rural areas, i.e. creating local food markets, online platforms and on-farm activities educational practices initiated both by consumers' movements and farmers, both actors are able to restore a direct relationship between *consumers and farmers*, which is in favour of both parties, where the trust and better understanding of each other plays the major role.

By deepening the agro-food supply chain and by implementing different short food supply chains, such as face-to-face, spatial proximity and spatially extended cooperation customers movements enabled different and more trustworthy cooperation between *customer and food relationships*, where fair, high-quality, sustainable and/or organic food reach the customer. The deepening of the agro-food supply chain provides the platforms for conducting practice-oriented workshops for consumers and customers movements who are interested in alternative and high-value food and who are involved in food pilot projects, participating in demonstration farms, which finally creates and empowers the better and deeper relationships between customers and food (Fig. 4.2).

Finally, the regrounding the mobilization resources, when consumers' movements actively participate and seek direct cooperation and relationships with small farmers and small family agribusinesses creates direct relationships between rural and urban agents as well it does connects *rural and urban business relationship* and have an important effect on rural development.

Back-to-the-land movements. Back-to-the-land movements are one of the most interesting and most recent movements in the post-industrial movements' era. And even in the Western part of the world 'back-to-the-land' social movements became greatly popular in the late 1960s and early 1970s, supported by a mostly idealistic group of people who wanted to live life more simply (Jacob & Brinkerhoff, 1986); in Eastern part of Europe this movement was late for at least fifty years and only in 2000s and 2010s social movements 'back-to-the-land' arose and are still trying to get the public attention.

Post-industrial social movements, and the back-to-the-land movement as an explicit example, are challenging long-standing urban–rural people business dichotomy by promoting an integrated conception of cities and countryside based on their functional interdependencies and showing that earlier dichotomy and separation of urban and rural life are no longer relevant to the post-industrial reality of nowadays. So it is clear that the direction from rural-to-urban areas is no longer predominant but vice versa—more and more urban residents are looking for new livelihoods opportunities in rural areas, as well as their interests in rural areas are

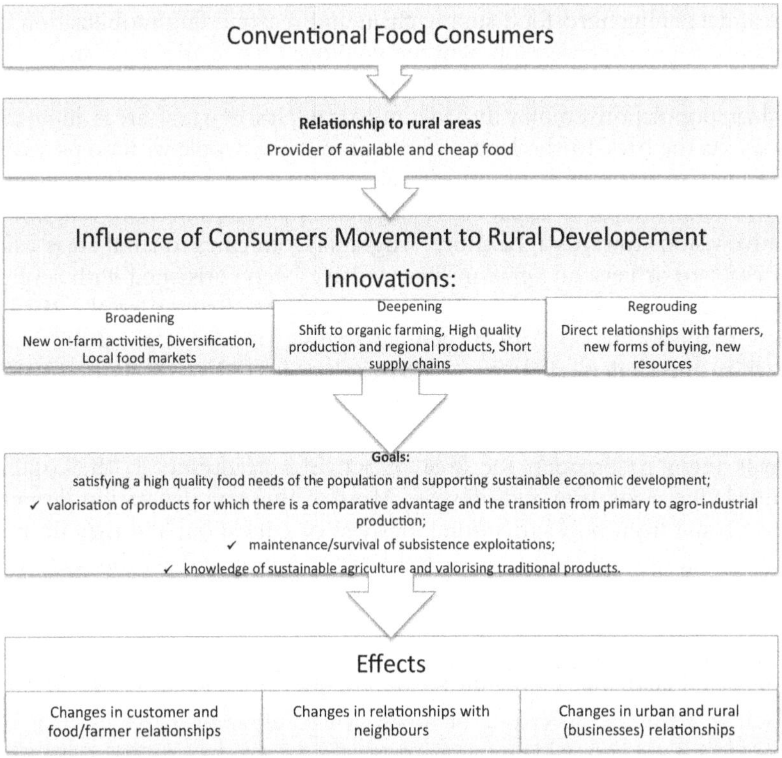

Fig. 4.2 Changing customers' goals and effects in industrial and post-industrial rural development (*Source* Developed by authors based on Van der Ploeg and D. Roep [2003])

not only limited to satisfying the food needs, economic development or employment in urban and rural areas separately.

Back-to-the-land movements are much more complex and may have many different reasons to be interested in rural areas. As it is mentioned in Chapter 5, Benessaiah (2018, p. 28) described at least eight motivations for going back to land, which are economic, social, political and cultural and which are interrelated with healthy lifestyle, reconnection to the nature and better quality of life. So again, by taking into account Van der Ploeg and Roep's (2003) research on three directions—broadening rural area, deepening agro-food supply chain and regrounding mobilization of resources—of rural development are employed for further analysis.

Broadening rural areas. Back-to-the-land movements as agents of urban population employ broadening strategies of rural areas in many ways. As the back-to-the-land movement involves people with no previous training or experience in agricultural activities, these new residents and farmers of rural areas usually have different, more autonomous and innovative visions towards agriculture. Migrants from cities to rural areas who attempt to achieve an agrarian lifestyle have been christened with several labels: neofarmers (Mailfert, 2007), neopeasants (Brunori et al., 2013), new pioneers (Jacob, 1997), new agrarians (Tregear et al., 2007) and back-to-the-landers (Belasco, 2005) (as cited in Orria & Luise, 2017, p. 128). Most of them promote independence, interpersonal relationships, and use service-driven business model. Servitization of farming helps them to broaden the area of activities according to their skills, knowledge, avocation and lifestyle. Most of the new generation farmers have come from non-agricultural lifestyles or education, and they do not seek to adopt agriculture as a full-time vocation and in this case the servitization gives an opportunity to integrate their previous skills and knowledge into farming by developing gastronomy, ecological and transformative tourism, healing, amusement and sport services closely interconnected with their agricultural production (Vidickiene et al., 2019). As it is seen the spectrum of areas, where agrarian activities may be broadened are very wide in rural areas.

Deepening the agro-food supply chain. The main area, where the back-to-the-land movement deepens the agro-food supply chain, is food quality and its locality. Back-to-the-land movements are very important agents in empowering short food supply chains, as personal relationship, trust, face-to-face contact and knowledge about the product are very important aspects for both—rural and urban population and for further development of rural and urban businesses relationships, where both parties can benefit from high-quality food, both are empowering sustainable agriculture and both are valorising traditional products (Fig. 4.3).

Regrounding the mobilization of resources. There can be many different resources in back-to-the-land movements, where the most important are human resources, such as labour, time, knowledge, skills, community, cooperation and many others. As many of these resources are limited, back-to-the-land movements, the same as consumers' movements

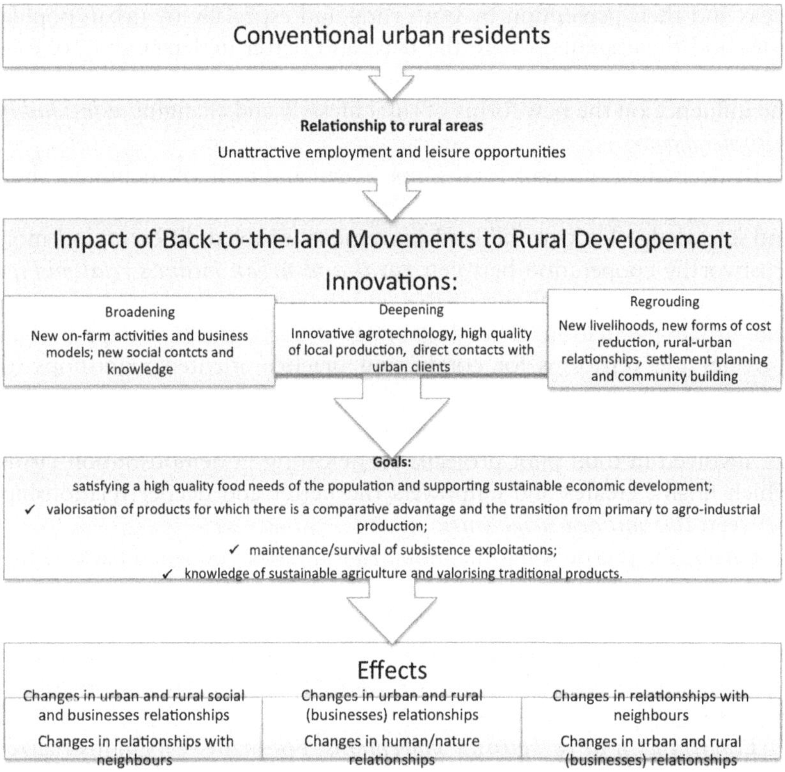

Fig. 4.3 Changing urban residents' goals and effects in industrial and post-industrial rural development (*Source* Developed by authors based on Van der Ploeg and D. Roep [2003])

have to be inventive and employ new forms of work organization, employ online communication, online markets.

Finally, there is a question, how these three strategies of broadening, deepening and regrounding strategies implemented by back-to-the-land movements are affecting rural areas and the shift of rural development paradigm? By broadening rural areas, i.e. developing gastronomy, ecological and transformative tourism, healing, amusement back-to-the-land movements creates new livelihoods which are changing ***rural and urban business relationships*** as well as it does change the appearance of rural areas and their perception by both rural and especially by urban populations and their agents, where the trust and better understanding of each other plays the major role. All the above-mentioned aspects as well have the influence on the new forms of rural lifestyle and changing *relationships with neighbours*.

By deepening the agro-food supply chain and by implementing a short food supply chains, a personal relationship, trust, face-to-face contact and knowledge back-to-the-land movements enabled different and more trustworthy cooperation between *rural and urban business relationships*, where fair, high-quality, sustainable and/or organic food reach the rural and urban population. The deepening of the agro-food supply chain provides the platforms for conducting practice-oriented workshops for back-to-the-land movements and other agents of both rural and urban population, who are interested in alternative and high-value food and who are involved in food pilot projects, participating in demonstration farms, which finally, creates and empowers the better and deeper relationships between *like-minded neighbours*.

Finally, the regrounding the mobilization resources, when back-to-the-land movements actively participate and seek for a direct cooperation and relationships with small farmers and small family agribusinesses creates direct relationships between rural and urban agents as well it does connects *rural and urban business relationship* and have an important effect on rural development.

Ecovillages and co-housing movements. For many years industrialized urban lifestyle had much more to offer to urban population, where all the aspects of industrial life were able to be enjoyed, however, as the 'green' world ideas become more popular, more and more people are looking for more sustainable or alternative lifestyle. Because of that, the so-called ecovillages, co-housing and other intentional communities, as well as various local employment and trading systems are getting more and more attention from urban population and priorities which have been important earlier, such as e.g. closeness to the workplace are no longer relevant. In the post-industrial stage of the society's development, the place of residence and the place intended for receiving income often no longer coincide (Vidickienė & Gedminaitė-Raudonė, 2011).

In order to realize the vision of an ideal settlement adapted to the features of post-industrial society, ecovillages began to develop at the end of the twentieth century. Soon, individual successful experiments gave rise to the ecovillage movement. The ecovillage movement is expanding, with international and global cooperation networks sharing their experiences. One of the largest networks in Europe connecting ecovillages is the 1995 The Global Ecovillage Network (GEN) which was set up to bring together ecovillage initiatives to create a sustainable natural environment, with a strong focus on social networking (Felicie, 2012; Global Ecovillage Network Europe, 2020). Ecovillages and co-housing movements can be described as an initiative looking for a more sustainable way of living in rural areas and promoting a lot of innovations which deal with climate change, environmental issues, resource shortages and social problems people face nowadays.

And even ecovillages and other co-housing movements can be very different, they are visible and vivid actors of rural development acting in three different but overlapping directions in broadening rural area, deepening agro-food supply chain and regrounding mobilization of resources (Van der Ploeg & Roep, 2003).

Broadening rural areas. Ecovillages and co-housing movements are broadening the rural areas the understanding and even the perception of rural areas in many ways. As Vidickienė et al. (2016, p. 54) described, there are at least three dimensions of the vision of ecovillages—ecological, social and spiritual. The vision of the spiritual dimension of the ecovillage must reflect the core spiritual values of the founders. Many ecovillages describe the spiritual dimension of their vision in terms, expressing the goal of living in harmony with the Earth and all the living creatures. Some ecovillages highlight their spiritual values on the basis of some concept of common understanding of the world described in philosophical theories. An important means of strengthening the spiritual life of the ecovillage is the cultural traditions created in the ecovillage, which promote the development and creativity of the inner world of the residents. The vision of the social dimension of the ecovillage defines the desired level of community and collectivity of the population. Some ecovillages seek to live like one big family, their residents share common resources, provide mutual assistance and engage in many joint activities. Other ecovillages prefer an individual lifestyle where the residents of ecovillage communicate like good neighbours, organize some traditional social events and meetings to discuss the development of the ecovillage, but do not seek to spend most of their time together. The vision of the ecological dimension of the ecovillage defines the main ecological values of the founders and the appropriate technical ways to implement them in decision-making on building architecture, building materials used, rural public infrastructure and the impact on other rural people, such as livestock farming, developing on-farm businesses. Ecovillages and other co-housing movements bring new on-farm activities, enables diversification and servitization as well as new nature and landscape management and most important—it creates new livelihoods. Ecovillages are creating low-impact, environmentally harmonious living situations, implementing nature-friendly agriculture techniques as well as businesses and education centres.

Deepening the agro-food supply chain. As the name of 'ecovillages' says—the 'eco' part is one the most important aspects of ecovillages where sustainable lifestyle and short food supply chain are essential. Usually the natural resources found in the area are the main material from which the community infrastructure must be built. For example, young families build their homes in an ancient way, using cheap ecological materials such as clay, straw, sawdust, hemp fibre and many other natural goods found in the area or plants grown specifically for this purpose. In this way each member of community learns to become a friend of nature, and not an exploiter. Organic farming promotes the cultivation of local traditional crops, such as hemp, suitable for construction, household, food and medicine. Various animals (cows, goats, sheep) of the farm and birds contribute to the healthy diet of the community members. Communities are also looking for more environmentally friendly ways to provide the energy they need for everyday life, for example, in developing the equipment for extracting and using solar energy in the community (Vidickienė et al., 2016). So ecovillages and co-housing movements demonstrate that the areas where organic and sustainable resources are used are much more wider than in any other social movement or initiative, as labels 'ecological' and 'local' are used not only for food but for all aspects of social movement's life and moreover it does create even a new social order in rural communities.

Regrounding the mobilization of resources. Ecovillages are one the social movements, where the mobilization of resources might be regrounded to the biggest extent, as resources are closely related to the vision and values of ecovillages. That is why members of ecovillage not only valorize natural resources, but as well they try to use those natural resources which are the most sustainable, the closest and the friendliest to nature. Ecovillages are creating low-impact, environmentally harmonious living conditions, pioneering nature-friendly agriculture techniques where the infrastructure of community must be ecological.

Human resources are another set of resources essential to ecovillages, where labour, time, knowledge, skills, community, cooperation and many others play an important role in achieving the goals of the movement in creating new rural settlement infrastructure and spatial planning.

Finally, there is a question how these three strategies of broadening, deepening and regrounding are implemented by back-to-the-land movements and how they are affecting rural areas and the shift of rural development paradigm? The spiritual, social and ecological dimensions of ecovillages broaden the understanding of rural areas, as it is not seen as only agricultural area—even more ecovillages reorganize rural community life, it brings a new form of rural lifestyle which is changing *relationships with neighbours* as neighbours are understood as members of a big family. Ecovillages are advocating for a new type of living in rural areas, where rural life can be non-consumeristic, non-commodified and in favour of nature needs. Ecovillages and many other co-housing movements are broadening the understanding of rural areas by showing a more sustainable way of life, which is especially important in the spite of the climate crisis and resource shortages that the world faces (Fig. 4.4).

By deepening the agro-food supply chain and by implementing a short food supply chains, a personal relationship, trust, face-to-face contact and knowledge, back-to-the-land movements enabled different and more trustworthy *relationships between neighbours* where fair, high quality, sustainable and/or organic food and sustainable living conditions are important not only for the community but it does change the *relationship between urban and rural (businesses)* actors, as well the organic factor in ecovillages communities are changing the customer and food relationships. The deepening of the agro-food supply chain provides the experience of the benefits of organic food and new forms of rural lifestyle for ecovillages and other co-housing movements and other agents of both—rural and urban population, who are interested in organic, sustainable and high-value food and lifestyle overall, which finally, creates and empowers the better and deeper relationships between like-minded people.

Finally, the regrounding the mobilization resources, when ecovillages and co-housing movements have different attitudes towards natural resources and by all means prefer natural, organic and sustainable resources for their living conditions, create not only different, deeper relationships *between the customer and food (and other resources)*, but also it does changes *urban and rural (businesses) relationships*, as well as different

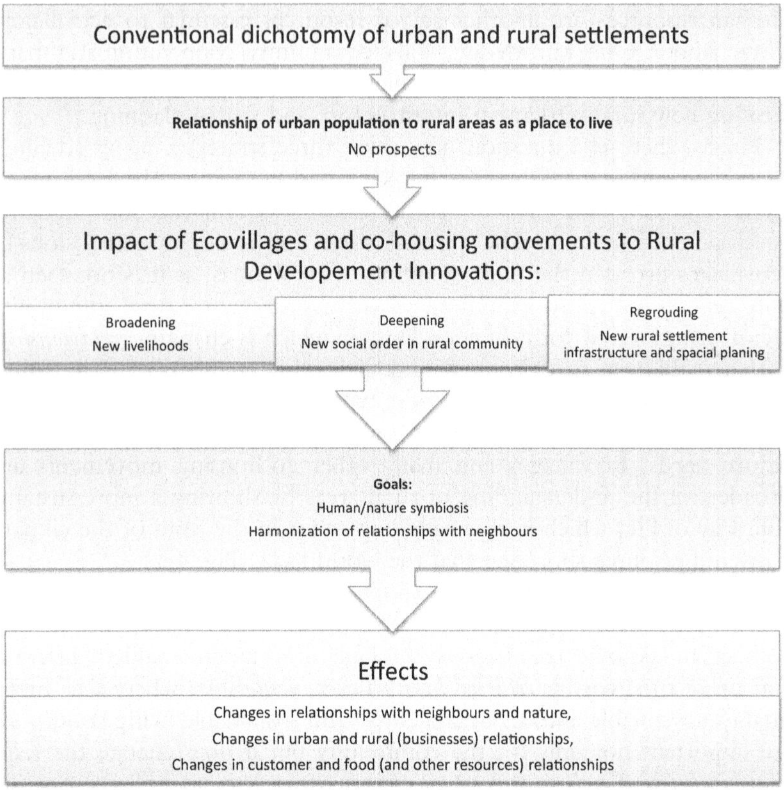

Fig. 4.4 Changing urban and rural settlements' goals and effects in industrial and post-industrial rural development (*Source* Developed by authors based on Van der Ploeg and D. Roep [2003])

organization of labour, leisure and other activities and does change the *relationships between neighbours.*

To sum up. To what direction the urban population is changing the rural development? Agents of urban population, such as consumers' movements, back-to-the-land movements and ecovillages/co-housing movements are very important aspects of rural development by broadening, deepening and regrounding rural development. All above-discussed actors of urban population participate in rural development shift in many different ways.

Consumers' movements, back-to-the-land movements and ecovillages by presenting new on-form activities, diversification, servitization, gastronomy and ecological and transformative tourism, healing, amusement, implementation of local food markets change the customer and food/farmer relationships. As well by bringing up the attention to the need of organic farming, sustainable living, high quality and locality of products and requiring for short food supply chains, consumers', back-to-the-land and co-housing movements are changing the relationships with neighbours and by helping in establishing direct relationships with farmers, implementing new forms of buying, they are changing urban and rural (business) relationships.

Consumers, back-to-the-land and intentional communities' movements not only changed conventional consumers', residents' and communities' attitudes from very simple ones, such as available and cheap food, unattractive employment and leisure opportunities, no prospects of relationships of urban population to rural areas as a place to live to much more complex—(1) satisfying high-quality food needs of the population and supporting sustainable economic development; (2) valorization of products for which there is a comparative advantage and the transition from primary to agro-industrial production; (3) knowledge of sustainable agriculture and valorizing traditional products. By reaching these goals, we can see many changes in urban and rural (business) relationships and changes in relationships with neighbours as well.

Finally comparing rural and urban population, we can see that both sets of actors are active in changing rural development, they are essential aspects of all the changes happening in both rural and even urban areas, which are related and connected to each other. And even sometimes it could look, that the primary goals of rural and urban agents are different, finally they do seek the same goals—sustainable rural development, rural welfare empowering social, economic and cultural development and the rural development which is not possible without non-technological drivers of progress, service-driven business model and especially collaborative and synergistic relationships between both—rural and urban population.

References

Agriculture Statistics—Family Farming in the EU. (2016). https://ec.europa.eu/eurostat/statistics-explained/index.php/Agriculture_statistics__family_farming_in_the_EU. 11 November 2020.

Baldock, D., & Beaufoy, G. (1993). *Nature conservation and new directions in the EC Common Agricultural Policy*. London: IEEP.

Belasco, W. (2005). Food and the counterculture: A story of bread and politics. In J. Watson & M. Caldwell (Eds.), *The cultural politics of food and eating* (pp. 217–234). Oxford: Blackwell.

Benessaiah, K. (2018). *Social-ecologies of crisis: Assessing the back-to-land movement in Greece*. A dissertation. Arizona State University.

Brunori, G., Malandrin, V., & Rossi, A. (2013). Trade-off or convergence? The role of food security in the evolution of food discourse in Italy. *Journal of Rural Studies, 29*, 19–29. https://doi.org/10.1016/j.jrurstud.2012.01.013.

Bryden, J. M., Bell, C., Gilliatt, I., Hawkins, E., & MacKinnon, N. (1992). *Farm household adjustment in Western Europe 1987-1991*. Final report on the research programme on farm structures and pluriactivity (Vols. 1 and 2, ATR/92/14). European Commission, Brussels.

Campbell, H. (2009). Breaking new ground in food regime theory: Corporate environmentalism, ecological feedbacks and the 'food from somewhere' regime? *Agriculture and Human Values, 26*(4), 309–319. https://doi.org/10.1007/s10460-009-9215-8.

Constance, D. H., & Renard, M. C. (2014). Alternative agrifood movements: Patterns of convergence and divergence. In M. G. Rivera-Ferre (Ed.), *Research in rural sociology and development* (Vol. 21, pp. 313–322). Emerald Group Publishing Limited. https://doi.org/10.1108/s1057-192220140000021007.

Felicie, A. (2012). *Global ecovillage network*. Newport: Salve Regina University.

Fernandez-Armesto, F. (2001). *Food: A history*. London: Pan.

Fuller, A., & Brun, A. (1991). *Farm family pluriactivity in Western Europe: Partial results of a research programme funded by the EC Commission and National Agencies*. The Arkleton Trust (Research) Limited: Oxford, UK.

Global Ecovillage Network Europe. (2020). https://gen-europe.org/. 16 November 2020.

Goodman, D. (2003). Editorial - The quality "turn" and alternative food practices: Reflections and agenda. *Journal of Rural Studies, 19*(1), 1–7. https://doi.org/10.1016/s0743-0167(02)00043-8.

Hassanein, N. (2003). Practicing food democracy: A pragmatic politics of transformation. *Journal of Rural Studies, 19*(1), 77–86. https://doi.org/10.1016/S0743-0167(02)00041-4.

Henderson, E. (2000). Rebuilding local food systems from the grassroots up. In F. Magdoff, J. Bellamy Foster, & F. H. Buttel (Eds.), *Hungry for profit: The agribusiness threat to farmers, food, and the environment* (pp. 175–188). New York: Monthly Review Press.

Jacob, J. (1997). *New pioneers: The back-to-the-land movement and the search for a sustainable future*. University Park: Pennsylvania State University Press.

Jacob, J., & Brinkerhoff, M. (1986). Alternative technology and part-time, semi-subsistence agriculture: A survey from the back-to-the-land movement. *Rural Sociology, 51*(1), 43–59. College Station, TX.

Knickel, K., & Renting, H. (2000). Methodological and conceptual issues in the study of multifunctionality and rural development. *Sociologia Ruralis, 40*(4), 512–528.

Mailfert, K. (2007). New farmers and networks: How beginning farmers build social connections in France. *Tijdschrift Voor Economische En Sociale Geografie, 98*(1), 21–31. https://doi.org/10.1111/j.1467-9663.2007.00373.x.

Makatouni, A. (2002). What motivates consumers to buy organic food in the UK? *British Food Journal, 104*, 345–352.

Marsden, T., Banks, J., & Bristow, G. (2000). Food supply chain approaches: Exploring their role in rural development. *Sociologia Ruralis, 40*, 424–438. https://doi.org/10.1111/1467-9523.00158.

Moore, O. (2006). Understanding post-organic fresh fruit and vegetable consumers at participatory farmers' markets in Ireland: Reflexivity, trust and social movements. *International Journal of Consumer Studies, 30*, 416–426. https://doi.org/10.1111/j.1470-6431.2006.00537.x.

Murdoch, J., Marsden, T. K., & Banks, J. (2000). Quality, nature and embeddedness: Some theoretical considerations in the context of the food sector. *Economic Geographyraphy, 76*, 107–126.

Orria, B., & Luise, V. (2017). Innovation in rural development: "neorural" farmers branding local quality of food and territory. *Italian Journal of Planning Practice, 7*(1).

Ortiz-Miranda, D., Moreno-Pérez, O. M., & Moragues-Faus, A. M. (2010). Innovative strategies of agricultural cooperatives in the framework of the new rural development paradigms: The case of the region of Valencia (Spain). *Environment and Planning A: Economy and Space, 42*(3), 661–677. https://doi.org/10.1068/a42168.

Renting, H., & Van der Ploeg, J. D. (2001). Reconnecting nature, farming and society: Environmental cooperatives in the Netherlands as institutional arrangement for creating coherence. *Journal of Environmental Policy & Planning, 3*, 85–101.

Tregear, A., Arfini, F., Belletti, G., & Marescotti, A. (2007). Regional foods and rural development: The role of product qualification. *Journal of Rural Studies, 23*(1), 12–22. https://doi.org/10.1016/j.jrurstud.2006.09.010.

Van der Ploeg, J. D., & Roep, D. (2003). Multifunctionality and rural development: The actual situation in Europe. In G. van Huylenbroeck & G. Durand

(Eds.), *Multifunctional agriculture; A new paradigm for European agriculture and rural development* (pp. 37–53). Hampshire, UK: Ashgate.

Vidickienė, D., & Gedminaitė-Raudonė, Ž. (2011). Lietuvos regionų klasifikavimas pagal kaimiškumą, remiantis atokumo kriterijumi. *Ekonomika Ir Vadyba: Aktualijos Ir Perspektyvos, 4*(24), 51–59.

Vidickienė, D., Gedminaitė-Raudonė, Ž., & Vilkė, R. (2016). Idealaus poindustrinės gyvenvietės modelio paieškos: ekokaimų atvejis. In *7-oji mokslinė-praktinė konferencija „Lietuvos kaimo vietovių konkurencingumo stiprinimas: geroji patirtis* (pp. 58–62).

Vidickienė, D., Gedminaitė-Raudonė, Ž., & Simonaitytė, V. (2019). *Ūkininkavimo servitizacija: verslo modelio „produktų gamyba plius paslaugų teikimas" apraiškos Lietuvoje.* Mokslo studija. Vilnius: Lietuvos agrarinės ekonomikos institutas.

CHAPTER 5

Rural Paradigm Shift and New Social Movements in the European Union

Vitalija Simonaitytė and Erika Ribašauskienė

5.1 Common Agricultural Policy (CAP) and New Rural Developments Trends

The Common Agricultural Policy (CAP) supports agriculture. Without the CAP, each Member State of the European Union (EU) would certainly implement its own national policies, and their scope and public intervention across various countries would differ. Commonly applicable policy provides general requirements for sustainable farming and provides common solutions and tools to address challenges related to market instability. It also helps to achieve greater competitiveness of the European agriculture, providing a basis for a Common Commercial Policy and thus enabling EU to be united when it negotiates with trading partners all over the world.

The main objectives of the CAP are the following:

- Protecting farmers from negative impacts of economic change and stabilizing their incomes;
- Climate change mitigation and sustainable management of natural resources;
- Fostering landscapes and ensuring the viability of the rural economy;
- Ensuring stable food supply chains.

R. Vilkė et al., *Rural Economic Developments and Social Movements*, Palgrave Advances in Bioeconomy: Economics and Policies,
https://doi.org/10.1007/978-3-030-71983-8_5

These objectives are common to all EU Member States and it is much easier to achieve them when the financial support for agriculture and rural areas is provided centrally in view of the major contemporary challenges.

5.1.1 *Development of the Common Agricultural Policy (CAP)*

The CAP emerged in the 1960s and has had a profound impact on the EU economy. Agriculture is a sensitive sector of economic activity, and its policies are constantly evolving in response to the sector's challenges and societal needs, bringing new reforms to make the CAP more effective.

The emergence of the CAP has been a relevant development due to specificity of agricultural sector and society's dependence on agricultural products.

As the population grew, so did the demand for food and, at the same time, its prices. This led farmers to intensify agricultural activities and increase production. The increase in food supply allowed to better meet the needs of the growing population, while the competition balanced agricultural output prices and slowed their growth. However, there was still a shortage of food in some European countries. Imports allowed people to better provide food at affordable price (Koning, 2006), however, the import taxes and also transportation costs grew resulting in higher food prices. Later, due to technological developments prices fell. The evolution of the chemical industry opened up opportunities for the use of cheap fertilizers and increased yields in European countries. With the emergence of agricultural production surpluses on the world market, prices fell sharply, farmers' incomes fell as well, and farm modernization was replaced by stagnation (Koning, 2006). Western European countries were forced to impose duties, in order to protect their own markets and farmers from surplus production in the international market (Koning, 2006).

After the Second World War, Europe faced food shortage, and all countries chose to protect their domestic markets. Agriculture was still the main 'employer' and relevant agricultural policies were needed to create conditions for producing the sufficient amount of food to satisfy domestic needs (Jambor & Harvey, 2010b). Discussions started regarding the integration of agricultural policy at the European level (Zobbe, 2002) and in the 1960s the CAP emerged. Its objectives were set out in the Treaty of Rome (Consolidated …, 1957):

- to increase agricultural productivity by promoting technological progress and ensuring the rational development of agricultural production and the optimal use of the factors of production, in particular labour;
- to ensure a decent standard of living for the rural population through these measures, in particular by increasing incomes of those working in agriculture;
- to stabilize markets;
- to guarantee food security;
- to ensure that products are supplied to consumers at reasonable prices.

The CAP has harmonized and consolidated the measures taken by the Member States into a single common policy, the principles of which are based on the following aspects (Konig, 2006): the single market; free internal trade, giving priority to EU products; equality and productivity: equalizing the incomes of farmers and workers in other sectors and reducing production prices by increasing productivity; co-financing: CAP expenditure is covered by the general budget collected from import taxes and other revenues.

Since the emergence of the CAP, a number of reforms have been implemented, which reflected the changing needs of farmers and society, increased the efficiency of the CAP measures and allowed to respond to the new challenges (Table 5.1).

The first significant changes to the CAP coincided with the reform of 1992, led by the EU Commissioner for Agriculture Ray MacSharry. The reform was influenced by the Uruguay Round Trade Negotiations and also driven by other external economic and social factors. Strong support for agricultural production through market regulation has led to surplus production and has shown that the CAP in its early years was so effective and attractive to producers that new problems arose. In order to increase the effectiveness of policies, it was decided to change the CAP's instruments by introducing coupled direct payments to farmers as a means of adapting to the reduction in funding for market regulation measures.

Public concerns for environment, food quality, and animal housing conditions necessitated new policies (Moehler, 2008). The reform of 1992 led to improvements in agriculture, but the European Commission (EC) was concerned that after 2000, surplus production would increase again due to export restrictions. The situation was also exacerbated by

Table 5.1 Development of the Common Agricultural Policy (CAP)

Productivity →

Competitiveness →

Sustainability →

Early years (1960)	Crisis years (1970-1980)	Reform of 1992	Agenda 2000	Reform of 2003	'Health check' in 2008
Food supply	Surplus production	Reduction of surplus production	Continuity of reform	Orientation to market demands	Strengthening of 2003 reform
Increase of productivity	Large expenses	Environment		Interests of consumers	
Market stabilization	International experience	Income stabilization	Competitiveness	Rural development	New challenges
				Environment	
Income stabilization	Structural measures	Budget stabilization	Rural development	Simplification	Risk management
				Alignment with WTO requirements	
In 1965 expenses for the CAP accounted for 35, 7 % of the EU budget, in 1985 these expenses grew and reached 70,8%, while in 1988–1992 financial year the CAP expenses accounted for 60,7 % of the EU budget (European Commission, 2007).			*In 2009, 50 billion EUR from the total payment appropriation of the budget, i.e., 44,6% (European Commission, 2010b).*		
First pillar			Second pillar		

Source Modified by the authors according the documents of the European Commission

the prospects for the EU enlargement, as the budget for supporting the agricultural sector was limited, making the distribution of the budget to a growing number of EU Member States a major challenge.

Agenda 2000 continued the implementation of the CAP reform, but its direction changed significantly, as the policy became focused not only on the welfare of farmers, but also on the rural population, consumers of agricultural products and the environment. It was Agenda 2000 that legitimized Pillar II of the CAP, envisaging the implementation of rural development measures and promotion of multifunctional agricultural activities (Jambor & Harvey, 2010a; 2010b). The EU rural development policy was initially sectoral and territorial, but Agenda 2000 harmonized the management of agricultural sectors across the EU (Káposzta & Nagy, 2008).

Although Agenda 2000 covered the period of 2000–2006, following the mid-term evaluation of the programme's measures, a policy change decision was taken; a new CAP reform was launched. Its essential provision was to decouple payments from production. The aim was to minimize surplus production by reducing the role of the most market-distorting forms of support. The reform was intended to help the CAP encourage farmers to produce not on the basis of direct payments based on output, but on the basis of demand and profitability, thus increasing competitiveness. In addition, there was a desire to redistribute support in favour of small farms. In 2003 the reform was intended to simplify the CAP and align its measures with the requirements and standards of the World Trade Organization (WTO).

On November 20, 2007 the Communication of the European Commission on the preparation of the CAP 'Health Check' was adopted. The main purpose of this 'Health Check' was to assess the consequences of the CAP reform in the light of current challenges and to review its measures to ensure the effectiveness of the CAP by 2013. The 'Health Check' highlighted new challenges for agriculture and rural development, which proved to be particularly relevant in shaping the CAP beyond 2013.

5.1.2 Pillars of the Common Agricultural Policy (CAP), Their Motives and Objectives

The emergence of the first pillar of the CAP (Pillar I) was driven by the need to ensure the stable level of agricultural production on the European Community (EC) market. Market regulation measures such as export subsidies, import restrictions and high intervention buying prices have encouraged farmers to produce more. Financial support was provided to help farmers to better adapt to the changing economic and social environment. These measures accounted for a large share of the budget in the early years of the CAP, but they had a negative impact on world markets, not always reflected the interests of farmers, and also caused environmental problems (Gay et al., 2005) and quickly lost taxpayers' approval.

The Green movement started criticizing farmers for intensive agricultural activities that damage the natural environment. In response to public criticism, in 1988 a voluntary set-aside policy was introduced and in 1992 it became mandatory. The MacSharry Reform emphasized the role of farmers in protecting the landscape and natural environment and their contribution to the well-being of rural communities (Moehler, 2008). In 1988 the Council of the European Union had to address three important issues (Moehler, 2008): budgeting, support for agriculture and structural policies to help regions experiencing difficulties to adapt to changing economic and social conditions. The lack of budgetary resources led to limited expenditure on price regulation. In order to better manage the CAP budget expenditure, so-called 'stabilizers'—production quotas—were introduced. Another problem was the low income of farmers, as most of the support went to exporters, traders and other intermediaries. R. MacSharry considered it appropriate to directly support farmers' incomes rather than regulating market prices (Moehler, 2008). He was able to persuade the Council of Ministers to change the form of support. In 1992 the support for cereal prices was changed to flat-rate area payments. To address the problem of agricultural surplus, the U.S. experience was taken into consideration (Jambor & Harvey, 2010b) and the set-aside requirement was introduced.

In 1997 the report of the expert group entitled 'Towards a Common Agricultural and Rural Policy' was published. It highlighted the main problems of the CAP and proposed guidelines for its further development. Experts identified four main elements of the European Common Agricultural and Rural Policy that focused on the problems of the time (Buckwell et al., 1997): ensuring market stabilization; environmental protection and landscape conservation; promoting rural development initiatives; support measures for farmers to adapt to new economic conditions during the transition period. It was clear that the CAP should take into consideration the factors of international competitiveness and global environment. The EU has become the world's largest player in the food market, with a free internal market (Buckwell et al., 1997), and farmers have become entrepreneurs with state support but little respect for the environment.

After the MacShary Reform in 1992 and reforms introduced by Agenda 2000, the CAP ensured adequate food security and farmers' incomes. However, the support provided encouraged farmers to produce more, leading to some surpluses.

In this context, the formulation of rural development measures under the Second Pillar (Pillar II) of the CAP became particularly relevant as the public began to express its views on the need for a separate policy to address the sensitive issues of rural development. The early rural development policy was sectoral (strongly linked to the structure of agriculture), but at the beginning of the twenty-first century, the rural policy started emphasizing cultural rather than economic principles, giving priority to the well-being of future generations. Agenda 2000 separated the Pillar

II of the CAP—Rural Development Policy—and harmonized its governance across the EU (Káposzta & Nagy, 2008). Since 2003, the European Commission has sought to allocate a larger share of the budget to Pillar II, expressing growing concern about the irrational waste of natural resources and environmental pollution. National (sometimes regional) development programmes have been put in place to meet the specific needs and challenges of rural areas. Although individual EU countries envisage similar support measures in their programmes as other countries, they have some flexibility to use them to address their major challenges, taking into account the specific economic, natural and structural conditions of specific areas. An integral part of rural development programmes is the LEADER approach, which encourages local people to address issues of local communities.

As the CAP has evolved, so has the distribution of its budget between Pillar I, which supports agricultural producers, and Pillar II, which supports rural development. In 1980, the bulk of the CAP expenditure went to support agricultural prices (export subsidies and other market regulation measures), and in the late 1980s, more funding went to market regulation measures to reduce agricultural surpluses. In 1992, the MacSharry Reform changed the structure of the CAP spending: spending on market regulation measures was reduced and redirected to support other measures coupled to direct payments. Thus, the maintenance of agricultural production prices was replaced by direct support to the producer. Expenditure in support of rural development increased in line with the growing share of direct payments. As a result of the 2003 Reform, direct payments were decoupled, while the funding for rural development measures increased. Under the Multiannual Financial Framework 2014–2020 (the EU's seven-year spending plan), 38% of total funds is allocated to the CAP budget, that is, a total of 408.31 billion EUR. 308.73 billion EUR is allocated for direct payments and market measures, while 99.58 billion EUR is allocated for rural development.

5.1.3 Rural Development and Its New Trends

The rural development paradigm was born in the 1960s as a separate academic discipline and underwent many transformations. New ideas of rural development were proposed by theorists and implemented in practice as a part of rural development policy measures in many countries of the world (Vidickiene & Melnikiene, 2014).

The concept of rural development has evolved and its features have changed during each decade and each programmatic period (Table 5.2).

In the last decade of the twentieth century, integrated rural development was seen as promoting change and creating more favourable conditions for rural human development by encouraging farmers to increase production and efficiently provide surplus production to cities (Ellis & Biggs, 2001). Such former rural development policies (especially in the period of 1960–1980) were intended to encourage the modernization of agriculture (especially in developing countries).

In 1990, the new dimension of sustainable development emerged in rural development approaches (Brundtland Report, 1987), according to which sustainable development is the development that meets the current

Table 5.2 The evolution of the concept of rural development implemented within the framework of the EU Common Agricultural Policy (CAP)

2000	2010	2020	2030
▪ Balanced income sources ▪ Good governance and decentralization, participatory approach ▪ Broad multi- sectorial approach ▪ Social protection ▪ Poverty reduction			
	▪ Strengthening of local identity ▪ Creation of welfare ▪ Diversification of production and activities ▪ Strengthening of social cohesion ▪ Application of innovative solutions ▪ Production and management of public goods ▪ Promotion of sustainability ▪ Creation of cultural environment and related services ▪ Development of local markets and local consumption		
		▪ Neo endogenous development and increased efforts to ensure sustainability ▪ Multi-functional agriculture ▪ „Bottom-up" approach ▪ Interterritorial and international collaboration ▪ Promotion of networks ▪ Priority for innovative projects contributing to sustainable development ▪ Local food systems and local interactions	

Source Modified according to J. Calatrova (2016) and F. Ellis and S. Biggs (2001)

needs of humanity, without compromising the ability of future generations to meet their own. It is a long-term approach for the continuous development of society, and the rational management of natural resources.

In 2000, the concept of rural development introduced a broad cross-sectoral approach to help to integrate activities and areas and facilitate the sustainable development of rural areas. The concept of the integrated rural development has the following features: a cross-sectoral approach based on local capital; rejection of the idea that agriculture is the sole engine of rural development (especially in developed countries); decentralized policy framework; growth to meet basic needs and achieve equal opportunities in developing countries; the importance of local solutions (involvement and social participation); development based on the use and strengthening of local, human and material resources (Calatrava, 2016).

In the second decade of the twenty-first century, the general objectives of rural development policies have been formulated by countries with well-developed agricultural sector. Rural development policy does not directly aim to promote agricultural growth, but seeks to redefine the role of agriculture in society. The Cork 2.0 Declaration on a Better Life in Rural Areas (2016) notes that rural areas and communities have an important role to play in achieving the United Nations Sustainable Development Goals (SDGs), but should also play an active role in implementing the principles of the 21st United Nations Framework Convention on the Climate Change (UNFCCC). Participants in the 2nd European Conference on Rural Development held in Cork announced that an innovative, integrated and inclusive rural and agricultural policy in the European Union should be pursued in line with ten policies (CORK 2.0 Declaration, 2016, look Table 5.3).

Sustainability of rural development can be achieved by implementing all mentioned policy orientations and by coordinating and integrating project activities. Priority is given to forms of agricultural development that support and promote broad societal goals, including landscape valorization and biodiversity conservation, strengthening links between rural and urban areas, organization of short supply chains, and increasing employment opportunities in rural areas.

In 2017, the European Commission published a new Communication on the Future of Food and Farming, building on the recommendations of the Cork Declaration 2.0 on rural development. The Communication emphasized sustainable development, the conservation of natural resources and the need to ensure intergenerational renewal. With regard

Table 5.3 'A Better Life in Rural Areas'—Policy Orientations

Point 1:	**Promoting Rural Prosperity**
Point 2:	Strengthening Rural Value Chains
Point 3:	Investing in Rural Viability and Vitality
Point 4:	Preserving the Rural Environment
Point 5:	Managing Natural Resources
Point 6:	Encouraging Climate Action
Point 7:	Boosting Knowledge and Innovation
Point 8:	Enhancing Rural Governance
Point 9:	Advancing Policy Delivery and Simplification
Point 10:	Improving Performance and Accountability

Source https://enrd.ec.europa.eu/sites/enrd/files/cork-declaration_en.pdf

to the latter, the Communication calls on the Member States to draw up programmes that reflect the needs of young farmers and propose simplification of the arrangements for young farmers to use financial instruments to support investment in farms and working capital. In addition, the Communication sets out new priority areas, focusing on rural value chains in areas such as clean energy, the emerging bio economy, circular economy and eco-tourism.

The evolution of rural development policy reveals the complexity of political issues at hand. Modern rural policy focuses on the sustainable implementation of economic, social and environmental dimensions. The key question is how to ensure and increase the capacity of rural areas to use processes, resources and actions outside their territory in a way that maximizes their benefits. This is the essence of the neo-endogenous or new internal model of rural development. Within the framework of this model, the focus is put on the dynamic interactions between localities and the external political, institutional and economic environment and how these interactions are mediated (Ward et al., 2005). Therefore, rural development policy cannot exist without a network to support greater stakeholder involvement and continuous dialogue on the governance of this policy. Networking ensures the integration and coordination of different interests, a trade-off between them, and at the same time facilitates the integration of different sectors representing the economic, social and environmental dimensions. Due to these reasons, networking plays a very important role in the context of sustainable rural development, involving a very large number of actors in the rural development process at different levels within variety of socio-economic contexts, and having wide range of needs, priorities, interests and expectations. It can be said that the use of networking as an instrument for the implementation of sustainable rural development marks a transition towards a new way of thinking about the issues and trends of rural development and also marks the emergence of a new rural development model.

5.2 European Green Deal and New Rural Social Movements

On December 11, 2019, the European Commission pre-sented its ambitious new proposal for a European Green Deal (EGD). The European Green Deal is a response to climate and environmental-related challenges. It is a new growth strategy that aims to transform the EU into a fair and prosperous society, with a modern, resource-efficient and competitive economy where there are no net emissions of greenhouse gases in 2050 and where economic growth is decoupled from resource use. It also aims to protect, conserve and enhance the EU's natural capital, and protect the health and well-being of citizens from environment-related risks and impacts (COM(2019) 640 final). EGD consists of ten pillars: (1) climate ambition—'Climate neutral' Europe, where the EU aims to reach net-zero greenhouse gas emissions by 2050; (2) Clean, affordable and secure energy; (3) Clean and circular economy, which include a sustainable product policy in order to use less materials, and ensure products can be reused and recycled; (4) Sustainable and smart mobility; (5) Greening the Common Agricultural Policy/'Farm to Fork' Strategy which aims for a green and healthier agriculture system; (6) Preserving and protecting biodiversity which includes measures to tackle soil and water pollution as well as a new forest strategy; (7) A zero-pollution ambition for a toxic free environment; (8) Mainstreaming sustainability in all EU policies; (9) The EU as a global leader in continuing to lead the international climate and biodiversity negotiations, further strengthening the international policy framework; 10) a European Climate Pact. The absolute majority of EGD's future proposals and strategies have to be reached in 2020 and some of them until the middle of 2021 (Table 5.4).

Current EGD is a result of a long and ongoing debate on climate changes in Europe, where a very important role was and is played by

Table 5.4 EGD goals, targets and future strategies

EGD goals	*Targets and future strategies*
Climate ambition	Proposal on a European 'Climate Law' enshrining the 2050 climate neutrality objective Comprehensive plan to increase the EU 2030 climate target to at least 50% and towards 55% in a responsible way Proposals for revisions of relevant legislative measures to deliver on the increased climate ambition (following the review of Emissions Trading System Directive; Effort Sharing Regulation; Land use, land use change and forestry Regulation; Energy Efficiency Directive; Renewable Energy Directive; CO2 emissions performance standards for cars and vans) Proposal for a revision of the Energy Taxation Directive Proposal for a carbon border adjustment mechanism for selected sectors New EU Strategy on Adaptation to Climate Change
Clean, affordable and secure energy	Assessment of the final National Energy and Climate Plans Strategy for smart sector integration 'Renovation wave' initiative for the building sector Evaluation and review of the Trans-European Network—Energy Regulation Strategy on offshore wind
Industrial strategy for a clean and circular economy	EU Industrial strategy Circular Economy Action Plan, including a sustainable products initiative and particular focus on resource intense sectors such as textiles, construction, electronics and plastics Initiatives to stimulate lead markets for climate neutral and circular products in energy intensive industrial sectors Proposal to support zero carbon steel-making processes by 2030 Legislation on batteries in support of the Strategic Action Plan on Batteries and the circular economy Propose legislative waste reforms

EGD goals	*Targets and future strategies*
Sustainable and smart mobility	Strategy for sustainable and smart mobility Funding call to support the deployment of public recharging and refueling points as part of alternative fuel infrastructure Assessment of legislative options to boost the production and supply of sustainable alternative fuels for the different transport modes Revised proposal for a Directive on Combined Transport Review of the Alternative Fuels Infrastructure Directive and the Trans European Network—Transport Regulation Initiatives to increase and better manage the capacity of railways and inland waterways Proposal for more stringent air pollutant emissions standards for combustion-engine vehicles
Greening the Common Agricultural Policy/'Farm to Fork' Strategy	Examination of the draft national strategic plans, with reference to the ambitions of the European Green Deal and the Farm to Fork Strategy 'Farm to Fork' Strategy Measures, including legislative, to significantly reduce the use and risk of chemical pesticides, as well as the use of fertilizers and antibiotics
Preserving and protecting biodiversity	EU Biodiversity Strategy for 2030 Measures to address the main drivers of biodiversity loss New EU Forest Strategy Measures to support deforestation-free value chains

(continued)

Table 5.4 (continued)

EGD goals	*Targets and future strategies*
Towards a zero-pollution ambition for a toxic free environment	Chemicals strategy for sustainability
	Zero pollution action plan for water, air and soil
	Revision of measures to address pollution from large industrial installations
Mainstreaming sustainability in all EU policies	Proposal for a Just Transition Mechanism, including a Just Transition Fund, and a Sustainable Europe Investment Plan
	Renewed sustainable finance strategy
	Review of the Non-Financial Reporting Directive
	Initiatives to screen and benchmark green budgeting practices of the Member States and of the EU
	Review of the relevant State aid guidelines, including the environment and energy State aid guidelines
	Align all new Commission initiatives in line with the objectives of the Green Deal and promote innovation
	Stakeholders to identify and remedy incoherent legislation that reduces the effectiveness in delivering the European Green Deal
	Integration of the Sustainable Development Goals in the European Semester
The EU as a global leader	EU to continue to lead the international climate and biodiversity negotiations, further strengthening the international policy framework
	Strengthen the EU's Green Deal Diplomacy in cooperation with Member States
	Bilateral efforts to induce partners to act and to ensure comparability of action and policies
	Green Agenda for the Western Balkans
Working together—a European Climate Pact	Launch of the European Climate Pact
	Proposal for an 8th Environmental Action Programme

Source Annex to The European Green Deal, COM (2019) 640 final, Brussels, 11 December 2019 social movements and public attitudes towards climate change on the one hand and political attitudes, Green political parties and political ideologies on the other.

Social background of EGD: social movements and public awareness on climate change. The environmental movement is commonly regarded as one of the more successful social movements of the last half of the twentieth century, in the United States as well as Europe (Dalton et al., 1990; Dunlap & Mertig, 1992). Although it can be argued that environmentalism has enjoyed only limited success in halting ecological deterioration (Dowie, 1995), very few social movements fully achieve their goals and most fail to survive for more than a few years (Mauss, 1975). In addition to the impressive staying power and large organizational base of environmentalism, the movement has clearly had significant institutional and cultural effects within most industrialized nations and beyond (Buttel, 1992; Dalton, 1994). A key reason for the success of environmentalism, relative to that of most social movements, is that its goal of environmental protection is widely supported by the general public (Hofrichter & Reif, 1990). Public support is a crucial resource for any social movement (Giugni, 1998, pp. 379–380), and the largely consensual nature of environmental protection has given the environmental movement an advantage over movements that pursue more divisive goals (Dunlap, 1995; Mertig & Dunlap, 1995; Mitchell, 1990). Indeed, of all the contemporary social movements, environmentalism is often seen as the one with the greatest level of actual and potential public support (Scott, 1990, as cited in Mertig & Dunlap, 2001, pp. 113–114). However, it has to be admitted that environmentalism movement laid an important and strong foundation for future environmental and climate change initiatives and decisions (Fig. 5.1).

Discussion on the environmentalism, climate change and social movements regarding these issues is impossible without mentioning the concept of Anthropocene. The concept of Anthropocene is trying to capture that human activity is having a dominating presence on multiple

Fig. 5.1 European Green Deal background in Europe

aspects of the natural world and the functioning of the Earth system, and that this has consequences for how people view and interact with the natural world—and perceive the place in it. As Malhi (2017, p. 78) discussed, the term has spilled across disciplines into the social sciences and humanities and into the wider cultural and political discussions surrounding how to live on and respond to the challenges of a human-dominated planet. Much of the vigour of this term now comes from these wider cultural and philosophical debates. The concept of Anthropocene includes emphasis on (a) the global and pervasive nature of the change; (b) the multifaceted nature of global change beyond just climate change, including biodiversity decline and species mixing across continents, alteration of global biogeochemical cycles and large-scale resource extraction and waste production; (c) the two-way interactions between humans and the rest of the natural world, such that there can be feedbacks at a planetary scale such as climate change; and (d) a sense of a current or imminent fundamental shift in the functioning of our planet as a whole (Malhi, 2017, pp. 78–79). And even the term of Anthropocene has not yet reached the public policy agenda directly, the influence of the concept and its goals is widely seen in many environmental policies, as well as in EGD. That is why it is intriguing to follow further developments of social movements and the impact of the concept of Anthropocene on their emergence, development and goal.

Seminal work on historical perspective of climate change has been made by Andrew Jamison in his book *The Making of Green Knowledge: Environmental Politics and Cultural Transformations*, where he described history of environmental ideas, importance of nature and knowledge as of Renaissance until 2000. However, the scope of this book and its starting point is since new social movements and climate change was first identified as a potentially significant public concern as one of the many aspects of an 'environmental crisis' that was to lead to the emergence of environmental movements in the 1970s. Like the other **social movements** that grew out of the student revolts of the 1960s—those of women's liberation and anti-imperialism, in particular the environmental movements, as they started to be called in the 1970s, were highly critical of the ways in which knowledge was produced in society (Jamison, 2010, p. 813). Jamison (2010, p. 813) describes social movements as 'as processes of political protest that mobilize human, material, and cultural resources in networks linking individual actors and organizations together in pursuit of a common cause. They provide spaces in the broader culture for new forms of knowledge-making and sociocultural learning as a central part of their activity'. The latter aspect of A. Jamison's definition of social movements is extremely important in the second so-called phase of environmental concerns developments—it is knowledge.

The emergence of **climate change knowledge** is strongly related to students movements in 1970s as universities and high-school students, and most of the activity was a collective learning in relation to environmental problems and dealing with what came to be termed 'the environmental crisis'. Activists and academics joined together to learn how to build solar energy panels and wind energy plants, grow organic food, and to try to live more ecologically—what we today would call climate-smart; i.e. finding ways to develop technology that do not emit carbon dioxide and other greenhouse gases. The environmental and energy movements of the 1970s also inspired the formulation of new ideas about science and technology, both for the production of energy but also more generally (Jamison, 2010, p. 814).

Above all 1980s can be identified as a backlash from environmental ideas to more conservative, nationally orientated problems and traditional values. At the same time as the anti-environmental 'backlash' was taking shape in the 1980s, the environmental movement itself fragmented into a number of different organizations and institutions, both in terms of politics and knowledge-making (Jamison, 2001). Green parties were formed in many countries and professional activist organizations, such as Greenpeace, grew in significance, while more broad-based grassroots organizations that led the campaigns against nuclear energy in the 1970s tended to weaken (Eyerman & Jamison, 1989). Universities and new environmental 'think tanks' started to make more specialized kinds of knowledge in areas such as renewable energy, organic agriculture and, eventually, in relation to climate change (Jamison, 1996, as cited in Jamison, 2010, p. 815). The idea of green business is based on a belief in a convergence between economic growth and environmental protection. Depending on the context, it has been termed ecological modernization, eco-efficiency, corporate sustainability or green growth (Jamison, 2010, p. 816).

The next step in environmental movements can be called **environmental and climate justice** movement. The global justice movement has been characterized as a 'movement of movements', a term coined by Naomi Klein in the wake of the anti-globalization protests of the late 1990s. The term captures well the heterogeneous character of the emerging submovement for climate justice, as well as the broader global justice movement (Klein, 2000).

The movement idea of climate justice originated with a focus on removing the causes of climate change, as well as addressing the inequitable impacts of the oil industry at all stages (from production and distribution through to climate impacts). But it also addressed fostering a 'just transition' to a post-carbon economy and providing assistance to vulnerable communities (Ikeme, 2003; Schrader-Frechette, 2002).

A recent discussion of the climate justice movement notes that it is based on principles of social justice, democratic accountability and participation, and ecological sustainability (Chatterton et al., 2013). These kinds of ideas, demands and principles can also be seen in the environmental justice movement, which has had a direct influence on the conceptualization of climate justice. The two thriving grassroots movements have influenced each other, and even fused in many ways. Both are at once international and local, and demand attention—and challenges—to the existing relationships between human communities and the environments that sustain them (Schlosberg & Collins, 2014, p. 370).

Environmental and climate justice is extremely important in the light of EGD, as EGD calls for 'leave no-one behind'. However at this point, it is not quite clear how this goal of 'no-one leaving behind' is going to be reached, but at least it shows that EGD has a way much broader perspective on climate change than it ever had been seen in the EU or even worldwide and it had not been possible without public concerns towards pollution, quality of people living environmental and climate change as a whole.

Attitudes towards climate change. In 2019, 23% of EU Members States' residents stated that climate change is the most serious problem facing the world as a whole, leaving international terrorism, the economic situation and armed conflicts far below (Special Eurobarometer, 490, 2019). However, it had not always been the case, e.g. the variable of climate change in measuring Europeans attitudes towards climate change was introduced only in 2004 in Special Eurobarometer. However, Fig. 5.2 shows a reliable trend that Europeans consider climate change a serious problem.

Because of different methodology, it is difficult to compare older data on Europeans attitudes towards climate change, but climate change has become a major concern for many citizens even earlier than in 2011 as

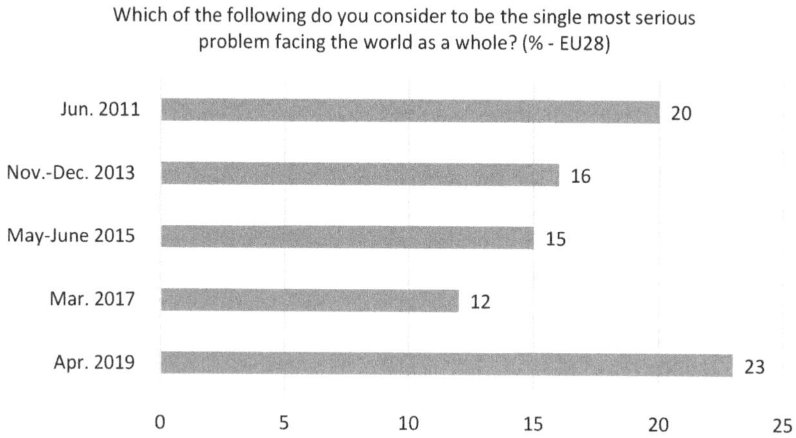

Fig. 5.2 Europeans' attitudes towards climate change (*Source* Special Eurobarometer 490, Report. Climate Change [2019])

Fig. 5.2 shows. For example, in 2008, when looking at the total aggregate of respondents' answers (analysis is based on all answers given by the respondents [i.e. first and other answers]) to what they consider to be the most serious problems,[1] 'global warming /climate change' rank second after 'poverty, the lack of food and drinking water'. While 68% think that 'poverty, the lack of food and drinking water' is one of the most serious problems our world faces now, 62% feel that 'global warming /climate change' also is among the most serious problems. At the country level, absolute majorities in nearly all countries regard global 'warming/climate change' as a serious problem, with the exception of citizens in the Czech Republic (45% consider this to be a serious problem), Italy and Portugal (both 47%). In Cyprus (92%) and Greece (90%) around nine in ten citizens think that 'global warming / climate change' is one of the most serious problems, in Slovenia this figure is as high as eight respondents in ten (Special Eurobarometer, 300). So, it is quite clear that climate change is seen as more and more significant problem in EU and in the world.

Changes towards political attitudes: Greens and environmentalism on European (Parliament) agenda. As C. Rootes stated 'There can be

[1] QE1 In your opinion, which of the following do you consider to be the most serious problem currently facing the world as a whole? Firstly? Any others?

no doubt that the development of the electorate for Green parties has been built upon unprecedented awareness of environmental problems. However much Green party theorists and activists insist on the distinction between environmentalism and ecologism, and most people in even the most environmentally aware European societies identify Green parties primarily with concern for the environment in the loose sense of opposition to pollution and environmental degradation. But, if consciousness of environmental deterioration is a necessary condition of support for Green parties, it is by no means a sufficient one; there is no simple correspondence between the state of environmental consciousness in a country and the level of development or electoral fortunes of its Green party' (Rootes, 1995, p. 170) (Fig. 5.3).

The results of European Parliament elections show that Green ideology and Green political parties play important role in political arena and their role is increasing; however, it cannot be stated that only the Group of the

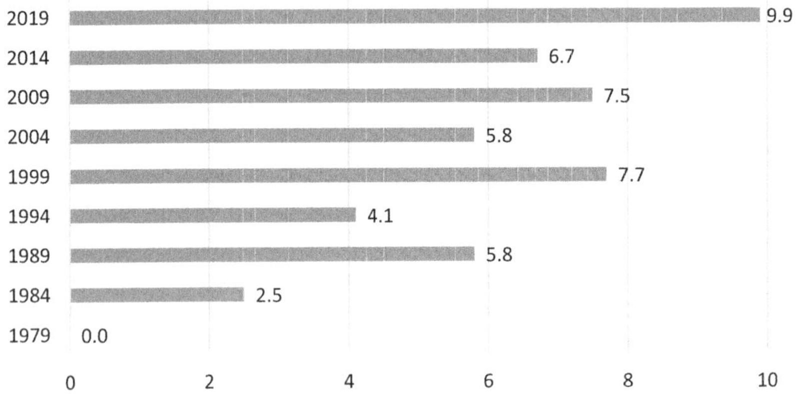

Fig. 5.3 Greens in European Parliament 1979–2019 (*Note* 2024–1999 Greens/EFA—Group of the Greens/European Free Alliance; 1999–1989 The Green Group in the European Parliament; 1989–1984 Rainbow Group [Federation of the Green-Alternative European Links, Agelev-Ecolo, the Danish People's Movement against Membership of the European Community and the European Free Alliance in the European Parliament] Greens held 11 seats in EP; 1984–1979 no Greens group in European Parliament. *Source* European Elections Results. https://www.europarl.europa.eu/about-parliament/en/in-the-past/previous-elections)

Greens/European Free Alliance made it happen. In particular in 2019 the society played extremely important role in bringing climate change issues to the public and to political actors. That is why it cannot be said that politicians made Green Deal happen, but more public pressure and politicians' willingness made Green Deal happen.

European Union's Environmental Policy. From its beginnings until 1985, environmental protection was neither included in the European Treaties nor defined in primary legislation as a European task. The EU expanded its responsibility for environmental questions, however, by a strategy known as 'frame bridging' (Snow et al., 1986) (Fig. 5.4).

In the preamble of the Treaty of Rome, the EU states its objective to improve life and employment conditions for its citizens. The Treaty's creators intended for the term 'life conditions' to be viewed in a strict economic light. However, the 'frame bridging' strategy enabled EU institutions to include, step by step, ecological 'living standards' as a relevant mission (Johnson & Corcelle, 1989; Knill, 2003, p. 19). With the Single European Act of 1987, the 'Treaty for the European Economic Community' expanded and separated environmental policy from other fields. Consequently, environmental policy was given its own Directorate

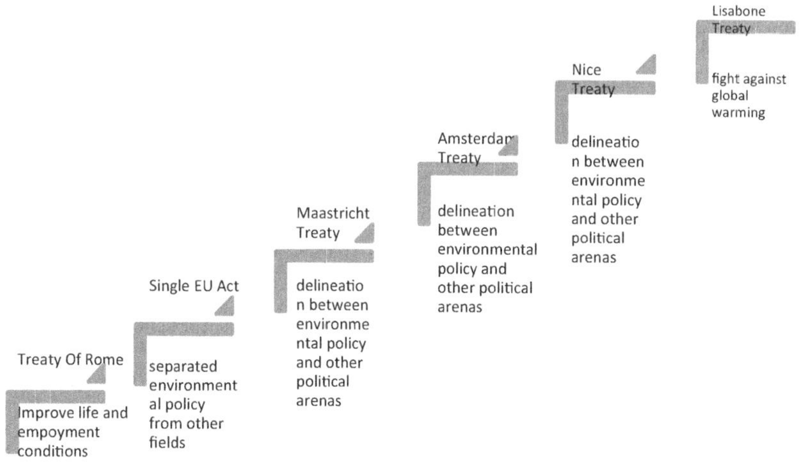

Fig. 5.4 Historical development of environmental policies in EU treaties until EGD (*Source* EU Treaties and Gerhards and Lengfeld [2008, p. 4])

General, which served to underscore the important institutional position of the environment. The Maastricht and Amsterdam Treaties (1993 and 1999, respectively) further strengthened this delineation between environmental policy and other political arenas. The Treaty of Nice mentions the importance of environmental policy as 'European Union play a leading role in promoting environmental protection' but it does not provide any details on how to reach this goal. The symbolic culmination of these institutional developments is the 'Reform Treaty of Lisbon' 2007. The Reform Treaty incorporates additional agreements regarding climate change and the fight against global warming, which have been added as targets for the European Union (Gerhards & Lengfeld, 2008, p. 4).

In light of these historical legal changes, it is clear that EGD is a huge or even 'first man on the moon' step for environmental and climate change policy but nonetheless, it is important how it will be implemented. That it is why not only top-down approach but also **bottom-up initiatives** are extremely important. Many local level initiators have stressed out that EU Green deal must be implemented not only by top-down but also by bottom-up approach, such as smart specialization, smart villages, etc. Even the history of new social movements showed that climate and environmental action can only be done through a bottom-up approach, and support for the participation of local communities. Smart rural communities are already providing many inspiring examples of how to address each of the challenges identified in the European Green Deal at local level (Slee, 2019). A meaningful 'European Green Deal' should aim for a top-to-bottom, long-term mobilization of public and private efforts in favour of a just transition towards a climate- neutral and sustainable economy, leaving no one behind. It is vital that Mrs von der Leyen's plans are executed alongside a bold financial plan that will not be seen as 'old wine in new bottles' (Game Changer: Financing the EGD, 2019).

The future of EGD has to be focused on inclusive development where no one should be left behind and even more—mobilization of local communities and collaborative arrangements with local groups leaders, researchers and political actors must be implemented.

5.3 Top-Down and Bottom-Up Actors in European Rural Areas

The bottom-up approach allows the local community and local players to express their views and to help define the development course for their area in line with their own views, expectations and plans. The bottom-up approach is viewed as an opposite approach to the top-down where decisions are made at the highest level, by political authorities but not by public or local actors.

There are a lot of different bottom-up actors, players and initiatives representing new social movements and influencing agricultural, environmental, social, economic and other policies. These initiatives of new social movements differ in their organization level, formality, the scope of activities and finally by the drivers of their activities. The main actors of new social movements in rural areas are small and large farmers, agricultural companies and cooperatives, landowners, residents of rural areas, rural communities, various associations uniting farmers and residents of rural areas, local action groups, organized local initiatives, initiatives uniting farmers and scientists, such as European Innovations Partnerships (EIP), local action groups (LAGs) organized by the LEADER's initiative, various services providers in rural areas, even individual actors and finally consumers. The spectrum of top-down and bottom-up actors is tremendously wide, as well as the drivers—why they participate in various rural activities.

One of the most important initiatives changing European rural areas is the **LEADER** approach. At this point, the bottom-up approach means that local actors, i.e. local action groups (LAG's) participate in decision-making about the strategy and in the selection of the priorities to be pursued in their local area. The involvement of local actors includes the population at large, economic and social interest groups and representative public and private institutions. LEADER conceives the local people as the best experts on the development of their territory. It can be seen as a **participatory democracy** tool supplementing the **electoral parliamentary democracy** (European Network for Rural Development).

The community initiative for rural development LEADER, launched in 1991, was an advanced initiative to promote new relations between public authorities and civil society, organized as local partnerships (Local Action Groups) in rural areas of Europe. This was a new form of governance, where different actors of the same territory gained legitimacy to access financial resources and act together in public life. Since 1991, until the previous programming period, the number of Local Action Groups, the territory covered and the financial allocation to LEADER have always increased from one programming period to another programming period. LEADER is a local development method, which has been used for 20 years to engage local actors in the design and delivery of strategies, decision-making and resource allocation for the development of their rural areas (LEADER/CLLD, ENRD).

Support for LAGs gradually increased from the very beginning of LEADER. At the experimental phase LEADER I (1991–1993) supported 217 LAGS with EU funding of 1.2 billion euro total public budget. The following LEADER period 1994–2006 was organized as Standalone LEADER programmes. LEADER II (1994–1999) focused on disadvantaged rural areas and supported 906 LAGs in 2017 regions by 5.4 billion euros total public budget. LEADER+ (2000–2006) initiatives covered all types of rural areas and supported 1 153 LAGs with EU funding of 5.1 billion euros total public budget. In a later period, LEADER was mainstreamed as an integral part of EU's rural development policy and supported 2 416 LAGs with funding of 8.9 billion Euro total public budget (LEADER 2007–2013 implementation update, 2015). In the period of 2014–2020, LEADER was extended into a broader Community-Led Local Development (CLLD) conception and added three more funds to utilize for bottom-up initiatives: the European Maritime and Fisheries Fund, the European Regional Development Fund and the European Social Fund (the European Commission, 2018). The ongoing programming period had entered its midway and collected data start giving evidence for on-going innovative bottom-up transformations of rural regions (Vilkė & Šarkutė, 2018, p. 512). It is implemented by around **2800 Local Action Groups (LAGs),** covering 61% of the rural population in the EU and bringing together public, private and civil society stakeholders in a particular area (situation as of end 2018—EU-28) (LEADER/CLLD, ENRD).

LEADER development approach fosters the 'bottom-up' principle of organization of communities, emphasizing the importance of self-organization and government's co-operation with the local rural population and how it is perceived from a new social movement theory standpoint (Vilkė & Šarkutė, 2018, p. 521). The main drivers of the LEADER approach are stronger, more organized, improved and cooperating rural regions of Europe. There are at least seven elements of the LEADER approach which contribute to qualitative local effects of the LEADER approach: directly addressing local issues and opportunities; strengthening stakeholder participation in local partnership and its governance; strengthening economic linkages among local actors; strengthening public–private partnership; mobilizing local resources; improving local community social capital and cohesion; improving local individual's knowledge, skills and capacities; finding and implementing innovative solutions to local problems; cooperating with other LAG territories (Table 5.5).

Regional development policy accelerated various bottom-up community demanded initiatives in a form of the LEADER programme. It

Table 5.5 Main characteristics and attributes of LEADER approach

Elements of the LEADER approach	*Qualitative local effects of LEADER approach*	*Objectives of Local Development Strategies*
Area-based local development strategies intended for well-identified sub-regional rural territories	Directly addressing local issues and opportunities	Knowledge transfer, education, capacity building
Local public-private partnerships (local action groups)	Strengthening stakeholder participation in local partnership and its governance	Climate change mitigation and adaptation
Bottom-up approach with decision-making power for local action groups concerning the elaboration and implementation of local development strategies	Strengthening economic linkages among local actors	Agriculture and farming, supply chains, local food
Multi-sectoral design and implementation of the strategy based on interaction between actors and projects of different sectors of the local economy	Strengthening public private partnership	Local economy (non-agriculture), job creation
Implementation of innovative approaches	Unpaid work carried out by LAG members	Culture, traditions, built environment
Implementation of cooperation projects	Mobilizing local/endogenous resources (human, physical, financial)	Natural environment and resources, landscape
Networking of local partnerships	Improving local community social capital and cohesion	Social inclusion, equality of opportunity, cohesion, services
	Improving local individual's knowledge, skills and capacities	Local governance and community development
	Finding/implementing innovative solutions to local problems	Broadband, internet, ICT
	Cooperating with other LAG territories	

Source Summarized by author based on European Network for Rural Development data assured financial resources to be mobilized for various bottom-up initiatives of local leaders in rural countryside. Therefore, local initiatives became provided with financial resources from the EU funds, which might be utilized in a form of registered local action groups (LAGs). Scientific research proved that accessibility to financial resources itself does not create sufficient background for new social movement to emerge; it should be necessarily aligned together with the motivation and skills of actors to utilize those resources creatively when moving rural development ideas forward, to become an organized bottom-up movement (Vilkė & Šarkutė, 2018, p. 522). However, it shows that LEADER is changing the rural areas, their development and overall the perception of rural areas and it is a great starting point and background for new social movement to emerge.

The European Leader Association for Rural Development (ELARD) is an international non-profit association set up to improve the quality of life in rural areas and to maintain their population through sustainable, integrated local development. The distinctive feature of ELARD is that it brings together Local Action Groups committed to involving all stakeholders in rural development at a local level. The European LEADER Association joins together almost 2200 Local Action Groups and Fisheries Local Action Groups from 26 countries, including from countries that are not members of the European Union but have adopted and are implementing the LEADER methodology (ELARD).

European innovation partnerships (EIP) is another European Union's initiative to bring together relevant parties at EU, national and regional levels to streamline, simplify and better coordinate existing financial instruments and initiatives. EIP 'Agricultural productivity and sustainability' launched in 2012, where main ideas were based on principles: 'more and better from less', 'ideas put into practice with success', impact by end users becoming motivated 'actors'. The EIP-AGRI applies an overarching '*Open innovation*' concept based on the interactive innovation model which is applied in EIP Operational Groups and H2020 Multi-Actor projects—collaboration between various actors to make best use of complementary types of knowledge (scientific, practical, organisational) in view of co-creation and diffusion of solutions/opportunities ready to implement in practice. European Union-wide EIP networks are linking actors in various activities: communication, partnering, dissemination, knowledge flows and collecting practice needs, which helps to generate '*Open science*' (EU SCAR AKIS, 2019).

EIP operational groups focus on challenges that can benefit society, modernize sectors and markets. They are launched in areas where combining EU, national and regional efforts in research and development will produce effective results. EIPs aim to coordinate investments in demonstration and pilots; anticipate and fast-track any necessary regulation and standards; better coordinate public procurement so breakthroughs are quickly brought to market (European Commission, EIP).

Even EIPs are administered by national authorities, but the emergence of the EIP operational group is a bottom-up initiative where at least three groups of actors—researchers, farmers and consultants come together and decide what problem they want to solve and what innovation they want to establish.

Even the main purpose of EIP is to bring researchers, farmers and consultants together to create and implement agricultural innovations, these operational groups have the much broader impact on their activities. First, they initiate and enable farmers, create networks and platforms between like-minded people of rural areas. Second, European innovation projects shape the understanding of other stakeholders, such as consultants, researchers, public policy institutions, of rural areas. In many cases the previous participation and experience in EIP encourages stakeholders to be more active and engaged in other initiatives, it creates bonds to other organizations, associations and people, participate in H2020, engage in activities of national associations, local action groups, etc. In this sense, EIP can be seen as bottom-up actors and their involvement in changing rural areas can be seen as the emergence of new social movements, as they shape the understanding of European rural areas and it brings a new focus on rural areas.

Local rural community associations are another important bottom-up actors in European rural areas. In Europe, there are a lot of European level organizations uniting stakeholders, which are related to EU rural development—European Rural Development Network (ERDN), The European Network for Rural Development (ENRD), European Rural Community Alliance's (ERCA) and many others. However, the scope of this research is not to analyse all of them, but to give a good example of how they are changing rural areas and their identity.

The European Network for Rural Development (ENRD) serves as a hub for exchange of information on how rural development policy, programmes, projects and other initiatives are working in practice and how they can be improved to achieve more. The ENRD is not a membership organization. Its work aims to engage and reach anyone with an interest in and commitment to rural development in Europe. The main stakeholders of the ENRD include National Rural Networks (NRNs); RDP Managing Authorities and Paying Agencies; Local Action Groups (LAGs); European organisations; Agricultural advisory services; Agricultural and rural researchers; and other interested rural development organizations and individuals. The ENRD supports the effective implementation of EU Member States' Rural Development Programmes (RDPs) by generating and sharing knowledge, as well as through facilitating information exchange and cooperation across rural Europe. Even the ENRD is a particularly important actor in European rural areas, however, it is not bottom-up actor par excellence, as it was established by initiatives of both—local actors and government institutions (ENRD).

In 2002 European Rural Development Network (**ERDN**) was established to integrate the efforts and competences of various European research institutions in their joint works on the state and paths of transformation of rural areas, in particular farming, with the view to the extension of the EU and its future policies. Thus, the main objectives of the ERDN are parallel to the community's idea of building the European research area for agriculture and rural development. The ERDN is meant to encompass the leading research centres studying rural development in Europe, and in particular in its central, eastern and south-eastern countries. The involvement of institutions, but mostly all of its individual members, makes it—apart from a research network—a forum for knowledge and information exchange based on mutual friendship, trust and goodwill. This decides on the unique character of the ERDN, where informal ties (social capital) and commitment of a group of people from various countries is the value added of the strictly scientific work. It is our achievement that we are proud of (ERDN).

European Rural Community Alliance's (**ERCA**) mission is to support the rural communities of Europe, through their national and regional rural movements, to develop connections, share experience and mutual learning and take collective action to strengthen their position. ERCA's role is focused on networking and information sharing, skill development and promoting the importance of the rural areas and their people (ERCA) (Table 5.6).

Table 5.6 European Rural Community Alliance's roles

Networking and information sharing	*Skills development*	*Promoting the importance of the rural areas and people*
Providing a platform for networking and exchange of experience, skills and knowledge between local communities and national networks Creating a pool of experience from rural communities across Europe on how to make the countryside viable and deal with rural issues Providing a platform for developing international connections and projects Enabling mutual support, solidarity and confidence in rural communities Spreading the idea and understanding of rural movements Providing an information for forthcoming events, reports, etc. Providing support for members in their engagement with policy makers	Providing a platform for mutual learning on the processes and structures of local rural development Providing learning opportunities between the well established and new/potential national rural movements. Developing community skills undertake their own research, dialogue with policy makers, project development and implementation Providing opportunities for local communities to experience, share and implement good practice from within and without the EU. Sharing and teaching practical tools and advice for village development	Raising awareness at national and EU levels about the extent, knowledge, experience and needs of rural communities Facilitating joint discussion between local communities and policy makers Enabling the voice of rural communities to reach decision makers at regional, national and EU levels Enabling local engagement in consultations on EU policies

Source Summarized by authors based on European Rural Community Alliance data

These three rural communities' associations represent European rural areas and their stakeholders. To a larger or smaller extent, they were initiated from political initiatives and are top-down movements. However recent developments and their activities represent that even the initiative to establish movements came from political level, movements are aggregating issues from their members, implement many activities and they are changing perception of rural areas and are working on many issues, such as rural, economic, social and environmental.

Another important set of actors in European rural areas are **small family farmers, agricultural companies and cooperatives**. One of the most recent initiative between farmers and agricultural companies is a so-called ***machinery ring***. A machinery ring is a cooperative society of farmers and agricultural businesses who have the mutual aim of reducing machinery and labour costs. The common objective for such community members is to reduce fixed and variable costs by collectively rationalizing the use of labour and machinery by sharing it in a formal manner. Members include farmers, hire companies, contractors, fuel/commodity suppliers and associated businesses connected with agriculture and the wider rural sector (e.g. Sastak machinery ring in United Kingdom) (Sastak Group). Machinery ring is a bottom-up initiative organized by farmers, contractors and other stakeholders reaching their goals. Machinery ring has dual effect on farmers and agricultural companies life: first of all, there are strong economic reasons to participate in machinery rings—farmers could improve their competitiveness and give adequate economic-social responses to the changing circumstances; secondly, there come social reasons for deeper cooperation—the community can offer support to the individual in case of personal crisis, and opens possibilities for the urban population (e.g. holidays, vacation) as natural part of their life. In addition to this, they participate in the organization and performance of community services in the local living environment (e.g. looking after public areas, joint marketing of rural tourism services, etc.) thus they reduce both the community and the individual financial expenses (Takács & Takács-György, 2012, p. 334).

Cooperative is another initiative changing the rural areas of Europe. Usually a cooperative is understood as an autonomous association of persons united to meet common economic, social and cultural goals. They achieve their objectives through a jointly owned and democratically controlled enterprise. European Commission promotes cooperatives' societies and focuses on 3 main issues: (1) the promotion of the greater use of cooperatives across Europe by improving the visibility, characteristics and understanding of the sector; (2) the further improvement of cooperative legislation in Europe; (3) the maintenance and improvement of cooperatives' place and contribution to community objectives (European Commission, Cooperatives). The importance of cooperation is obvious for the well-being of small and medium-sized farmers, consumers of agricultural products, as well as the state's economy and even for the social rural environment. The social and economic well-being of individuals and the state as a whole depends directly on the willingness and ability of business, the non-governmental sector and individuals to cooperate, as well as the ability of the public sector to support cooperation initiatives. As Igual and Melia-Marti (2008) point out cooperatives help maintain the social fabric and protect the environment. Their close relationship with their producer-members puts them in a privileged position for ensuring food safety and traceability while upholding their own cooperative values. These last turns them into inherently good corporate citizens.

Machinery rings and agricultural cooperatives both are bottom-up initiatives, established by farmers which are shaping the picture of European rural areas by implementing new social phenomena, such as sharing economy, digitalization of agricultural sector, even closer relationship with like-minded people, local communities and neighbors. Machinery rings and cooperatives are previously popular forms of social movements, which are now being replaced by movements pursuing more general goals, including the well-being of the rural and urban population, acting as international networks, and forming platforms for modern rural development.

Farmers. La Via Campesina International Peasant's Movement is a global, transnational movement fighting for food sovereignty, climate and environmental justice, international solidarity, agroecology and peasants' seeds, peasants' rights, land, water and territories, dignity for migrants and waged workers. LVC is a global alliance of organizations of family farmers, peasant farmers, indigenous people, landless peasants and farm workers, rural women and rural youth, representing at least 200 million families worldwide. In Europe, La Via Campesina unites 26 national organizations from 17 countries. Both agribusiness and rural social movements are attempting to re-territorialize spaces, that is, reconfigure them to favour their own interests, whether those are maximum extraction of profits or defending and building communities. A key aspect is that this involves not just a battle over land per se, but also very much a battle over ideas (Rosset & Martínez-Torres, 2012).

La Via Campesina is considered by many to be the most important transnational social movement in the world (Borras, 2004; Edelman, 2005; Martinez-Torres & Rosset, 2008; McMichael, 2006; Patel, 2005, 2006; Borras & Franco, 2009, as cited in Martinez-Torres & Rosset, 2010, p. 151). In contrast with other sectors—workers' unions, professionals, women, environmentalists, etc.—peasants and family farmers have been able to build a structured, representative and legitimate movement, with a common identity, that links social struggles on five continents. It has been identified as being among the grassroots movements that are 'the most innovative actors in setting agendas for political and social policies' (Yúdice, 1998, p. 353, as cited in Martínez-Torres & Rosset, 2010, p. 151). La Via Campesina can be seen as 'the international peasant movement', analogous to the 'international environmental movement', or 'the international women's movement', though Via Campesina has a tighter, more formal coordination than either of those two examples. It is also an autonomous and pluralistic movement (Martínez-Torres & Rosset, 2010, p. 150).

Another aspect, which has to be brought into light when analysing La Via Campesina is agroecology. Agroecology is a threefold phenomenon meaning the scientific discipline (plot/field approach, ecology of food system, agroecosystem ecology), movement (environmentalism, sustainable agriculture, rural development) and practice (technique). Agroecology encourages farmers and extensionists to participate in the design of new systems, and also contribute to social movements and there is often a link between a political vision (the movement), a technological application (the practices) to achieve the goals and a way to produce the knowledge (the science) (Wezel et al., 2009, p. 511). And even there is an increasing attention to the agroecology and the increasing number of agroecology movements (e.g. Agricultural and Rural Convention—ARC2020, European Association for Agroecology, etc.), social movements such as La Via Campesina are taking agroecology very seriously as agroecology is in their core: 'For many, agroecology is a science: the science that studies and attempts to explain the functioning of agroecosystems. For others, the word agroecology refers to the principles—not recipes—that guide the agronomic and productive practices that permit the production of food and fiber without agrochemicals... For the social movements that make up La Vía Campesina, the concept of agroecology goes much farther than just ecological-productive principles. In addition to these, LVC incorporates social, cultural and political principles and goals into its concept of agroecology' (Machín Sosa et al., 2010, p. 16, as cited in Rosset & Martínez-Torres, 2012, p. 17).

M. E. Martínez-Torres and P. Rosset (2010, p. 151) identified five phases in La Via Campesina's evolution of the movement. The first phase took place during the 1980s up to 1992, when several national rural movements felt the impact of similar global policies on local and national conditions. The second phase (1992–1999) was marked by the consolidation of continental networks in Latin America and the birth and structuring of La Via Campesina as a global movement. The third phase

(2000–2004) essentially consisted of becoming a key player on the international stage. The fourth phase (2004–2008) was marked by growth and internal strengthening, including setting up of regional secretariats, and the fifth, and current, phase (late 2008–present) responds to recent changes in the world and reflects a maturing political-economic analysis.

La Via Campesina is a bottom-up movement par excellence and it can be considered as new social movement as in many countries it brought new and relevant problems to public policy, as well as it helps to see the changing society and creates new identities of rural areas.

Other important actors in European rural areas are **consumers and residents of rural areas**. These actors usually seek better, more sustainable and environment friendly living conditions, better food and usually just different and better quality of life. Even there are many actors and bottom-up movements reaching these goals, a few of them are more recent and more changing the landscape of European rural areas—i.e. slow food movements and back-to-the-land movements.

One of the most recent and also one of the most important new social movements in Europe is a Slow Food International movement. The movement was established at the end of the 1980s in Italy and is one of the classic examples of new social movements. Slow Food is a global, grassroots organization seeking to prevent the disappearance of local food cultures and traditions, counteract the rise of fast life and combat people's dwindling interest in the food they eat, where it comes from and how the food choices affect the world around. Since its beginnings, Slow Food has grown into a global movement involving millions of people in over 160 countries, working to ensure everyone has access to good, clean and fair food. Movement Slow Food International believes that food is tied to many other aspects of life, including culture, politics, agriculture and the environment. Through the food choices, people can collectively influence how food is cultivated, produced and distributed, and change the world as a result (Slow Food International).

Slow Food International unites four networks: Terra Madre Network, Indigenous Terra Madre Network, Migrant Network and Slow Food Youth Network. Terra Madre is an international network of food communities—groups of small-scale producers and others united by the production of a particular food and closely linked to a geographic area. The network unites food producers, fishers, breeders, chefs, academics, young people, NGOs and representatives of local communities who are working to establish a system of good, clean and fair food from the grassroots level. The Indigenous Terra Madre (ITM) network seeks to bring indigenous peoples' voices to the forefront of the debate on food and culture and to institutionalize indigenous peoples' participation in the Slow Food movement, as an integral part of the larger Terra Madre network. Slow Food believes that defending biodiversity also means defending cultural diversity. The rights of indigenous peoples to control their land, to grow food and breed livestock, to hunt, fish and gather according to their own needs and decisions is fundamental in order to protect their livelihoods and defend the biodiversity (Slow Food International).

Since 2014, Migrant Network under Slow Food international is striving to promote the traditional knowledge of internal and international migrants from over 40 countries around the world. Slow Food Migrant network supports cooks, farmers, cheesemakers, beekeepers and many others who are cultivating their own diversity in their new homes, triggering gradual processes of adaptation and cultural and gastronomic cross-pollinations. The Slow Food Youth Network (SFYN) is a worldwide network of young people creating a better future through food. It unites groups of young food enthusiasts, chefs, activists, students and food producers who participate in public debate about current issues and introduce young people to the world of gastronomy. SFYN groups raise awareness about important food issues such as how to feed the world, food waste and sustainable food production, stimulating positive action (Slow Food International).

Even Slow Food movement is one of the best known, there are others initiatives and platforms such as URGENCI—Community Supported Agriculture and Forum Synergies, which are related to food movements in a much broader perspective. URGENCI promotes all forms of partnership between producers and local consumers, all kinds of Community Supported Agriculture initiatives, as a solution to the problems associated with global intensive agricultural production and distribution. Partnership, locality, solidarity and the producer/consumer tandem are the fundamental ideas of the movement. URGENCI stresses out that each producer–consumer partnership is independent, and it is based on direct person-to-person contact and trust, with no intermediaries or hierarchy and no subordination, also it has to be based on fairness, solidarity and reciprocity (URGENCI). Forum Synergies is an association of engaged citizens, organizations and active practitioners engaged in sustainable rural development. Forum Synergies focuses on civil dialogue between different stakeholders for sustainable rural development, rural youth, sustainable forestry practices by civil society, rural actors for health safeguarding traditional knowledge for the benefit of society, farmers ecology and agroecology, local ecological knowledge for future solutions, local food chains and sustainable development. Forum Synergies also developed a European map of rural sustainability, which covers success stories and organizations in 22 thematic areas, related to the goals of organization. So it is clear, that both movements are similar to Slow Food, as they are seeking for direct and fair relationship between farmers and consumers, as well as all of them stress the importance of local food sovereignty, fair, clean and safe food.

Movements 'back to the land' are one of the most interesting and most recent movements in the post-industrial movements' era. And even in the Western part of the world 'back to the land' social movements became greatly popular in post-Woodstock era in the late 1960s and early 1970s, supported by a mostly idealistic group of people who wanted to live life more simply (Jacob & Brinkerhoff, 1986); in many parts of nowadays Europe these movements are getting more and more support in response to economic, environmental or even more recent health crisis. This trend is seen in Greece as of the 2008 economic crisis, as well in many Western and Eastern European countries. K. Benessaiah (2018, p. 28) described at least 8 motivations for going back to land: (1) employment security: the need to secure a stable job and secure source of income; (2) new investment: primary sector seen as an investment opportunity for economic growth (without having a green growth focus); (3) green economy: desire to invest in new, more environmentally friendly, land production systems (geared towards creating a green business); (4) reconnect to the nature: need to be close to the environment and various attributes associated with living in more 'natural' areas; (5) being healthy: desire to eat 'good' food, without chemicals (usually organic), rediscovering the taste of food, also a general concern over health; (6) good life: search for a meaningful and good life, with better life and work conditions and having more time; (7) self-sufficiency: desire to become more autonomous, to get a sense of security and independence; (8) political action: people stating explicitly that they considered turning to land-based activities as a form of resistance and a political statement. It is clear that only a few of these motivations are economic, while the most of them are related to the social and environmental. This illustrates that rural areas are changing as well as the perception of rural areas. Even currently there are no organized back-to-the-land social movements at the European level, it is clear that these movements have a great potential for further and deeper development of European rural areas.

The European Rural Parliament (ERP) is another actor acting in the field of rural development, combining top-down and bottom-up approaches. ERP is a long-term campaign to express the voice of rural people in Europe, and to promote self-help and action by the rural people, in partnership with civil society and governments. The European Rural Parliament is designed to:

- Strengthen the voice of the rural communities of Europe, and ensure that the interests and well-being of these communities are strongly reflected in national and European policies.
- Promote self-help, common understanding, solidarity, exchange of good practice and cooperation among rural communities throughout Europe.
- Enable rural people to play their full role in addressing the great economic and political challenges which are currently affecting Europe (ERP data).

The European Rural Parliament is co-initiated by three pan-European networks: European Rural Community Alliance (ERCA), PREPARE Partnership for Rural Europe and European LEADER Association for Rural Development (ELARD). In addition to the three co-initiating partners, ERP also has European partners, which are Forum Synergies, European Council for the village and Small Town ECOVAST and ESIN—the European Small Islands Federation. So it is clear that ERP has a wide network to influence European agricultural policy and to affect rural development. An analysis of European top-down and bottom-up actors revealed that these actors, initiatives, organizations and movements focus on many overlapping problems and issues and no longer their main focus is only agricultural issues or exceptionally local areas. European top-down and bottom-up actors play important role in much broader sectors and issues, such as migrants and business, empowerment of women and youth, quality of life of rural residents and indigenous people, food and human rights. This trend clearly reveals the ties of discussed movements to new social movements and in many cases their development is similar to new social movement, as their already are changing the identity of rural areas and their perception.

References

Annex to The Communication from The Commission to The European Parliament, The European Council, The Council, The European Economic and Social Committee and The Committee of The Regions. The European Green Deal. Brussels, 11.12.2019. COM (2019) 640 final.

Benessaiah, K. (2018). *Social-ecologies of crisis: Assessing the back-to-land movement in Greece*. A dissertation. Arizona State University. https://repository.asu.edu/attachments/201334/content/Benessaiah_asu_0010E_17927.pdf.

Borras, Jr., S. (2004). *La Via Campesina: An evolving transnational social movement* (TNI Briefing Series no. 2004/6). Amsterdam: Transnational Institute.

Borras, Jr., S., & Franco, J. C. (2009). *Transnational agrarian movements struggling for land and citizenship rights* (IDS Working Paper no. 323). Brighton: Institute of Development Studies (IDS), University of Sussex.

Buckwell, A., et al. (1997). *Towards a common agricultural and rural policy for Europe: Report of an expert group* [online]. European Commission. Available from Internet. http://ec.europa.eu/agriculture/publi/buck_en/cover.htm.

Buttel, F. H. (1992). Environmentalization: Origins, processes, and implications for rural social change. *Rural Sociology, 57*, 1–27.

Calatrava, J. (2016, January 27–30). *Origin and evolution of rural development concept and policies: From rural communities to territories*. Old and New Worlds: The Global Challenges of Rural History. International Conference, Lisbon, ISCTE-IUL.

Chatterton, P., Featherstone, D., & Routledge, P. (2013). Articulating climate justice in Copenhagen: Antagonism, the commons, and solidarity. *Antipode, 45*, 602–620.

Consolidated Version of the Treaty Establishing the European Community Contents [online]. Roma, 25 of March, 1957. Available from Internet: http://eurlex.europa.eu/en/treaties/dat/12002E/htm/C_2002325EN.003301.html#anArt33.

CORK 2.0 Declaration 2016. A Better Life in Rural Areas. (2016). Available from Internet. https://ec.europa.eu/agriculture/sites/agriculture/files/events/2016/ruraldevelopment/cork-declaration-2-0_en.pdf.

Dalton, R. J. (1994). *The Green rainbow: Environmental groups in Western Europe*. New Haven: Yale University Press.

Dalton, R. J., Kuechler, M., & Bürklin, W. (1990). The challenge of new movements. In R. J. Dalton & M. Kuechler (Eds.), *Challenging the political order: New social and political movements in western democracies* (pp. 3–20). New York: Oxford University Press.

Dowie, M. (1995). *Losing ground: American environmentalism at the close of the twentieth century*. Cambridge: MIT Press.

Dunlap, R. E. (1995). Public opinion and environmental policy. In J. P. Lester (Ed.), *Environmental politics and policy* (2nd ed., pp. 63–114). Durham, NC: Duke University Press.

Dunlap, R. E., & Mertig, A. G. (Eds.). (1992). *American environmentalism: The U.S. environmental, 1970–1990*. Philadelphia: Taylor & Francis.

Edelman, M. (2005). Bringing the moral economy back in... to the study of 21st-century transnational peasant movements. *American Anthropologist, 107*(3), 331–345.

Ellis, F., & Biggs, S. (2001). Evolving themes in rural development - 1950s-2000s. *Development Policy Review, 19*(4), 437–448.

EU SCAR AKIS. (2019, October 10). *Preparing for future AKIS in Europe. Brussels, European Commission* (2nd ed.).

EU Treaties. https://europa.eu/european-union/law/treaties_en. 12 October 2020.

European Commission, Cooperatives. https://ec.europa.eu/growth/sectors/social-economy/cooperatives_en. 12 October 2020.

European Commission. European Innovation Partnerships (EIPs). https://ec.europa.eu/info/research-and-innovation/strategy/goals-research-and-innovation-policy/open-innovation-resources/european-innovation-partnerships-eips_en. [5 October 2020.

European Elections Results. https://www.europarl.europa.eu/about-parliament/en/in-the-past/previous-elections. 5 October 2020.

European Network for Rural Development (ENRD). https://enrd.ec.europa.eu/home-page_en. 5 October 2020.

European Network for Rural Development. Bottom-up Approach. https://enrd.ec.europa.eu/content/1-bottom-approach_en. 5 October 2020.

European Rural Community Alliance's (ERCA). http://www.ruralcommunities.eu/. 27 October 2020.

European Rural Development Network (ERDN). http://erdn.eu/. 27 October 2020.

Eyerman, R., & Jamison, A. (1989). Environmental knowledge as an organizational weapon: The case of Greenpeace. *Social Science Information, 2*, 99–119.

Forum Synergies. (2020). https://www.forum-synergies.eu/index_en.html. 27 October 2020.

Game Changer: Financing the European Green Deal. (2019). https://www.finance-watch.org/wp-content/uploads/2019/09/Financing-the-European-Green-Deal_Sept2019_03.pdf.

Gay, S. H., Osterburg, B., Baldock, D., & Zdanowicz, A. (2005). Recent evolution of the EU Common Agricultural Policy (CAP): State of play and environmental potential. Available from Internet. http://www.ieep.eu/publications/pdfs/meacap/WP6/WP6D4B_CAP.pdf.

Gerhards, J., & Lengfeld, H. (2008). The growing remit of the EU in environmental and climate change policy and citizens' support across the union. *SSRN Electronic Journal*. https://doi.org/10.2139/ssrn.2144672.

Giugni, M. G. (1998). Was it worth the effort? The outcomes and consequences of social movements. *Annual Review of Sociology, 98*, 371–393.

Hofrichter, J., & Reif, K. (1990). Evolution of environmental attitudes in the European Community. *Scandinavian Political Studies., 13*, 119–146.

Igual, J., & Melia-Marti, E. (2008). *Social economy and the cooperative movement in Europe: Contributions to a new vision of agriculture and rural development in the Europe of The 27* (pp. 147–172). CIRIEC-España, revista de economía pública, social y cooperativa.

Ikeme, J. (2003). Equity, environmental justice and sustainability: Incomplete approaches in climate change politics. *Global Environmental Change, 13*, 195–206.

Jacob, J., & Brinkerhoff, M. (1986). Alternative technology and part-time, semi-subsistence agriculture: A survey from the back-to-the-land movement. *Rural Sociology, 51*(1). College Station, TX, etc.

Jambor, A., & Harvey, D. (2010a). *CAP reform options: A challenge for analysis & synthesis* [online] (Centre for Rural Economy Discussion Paper Series No. 28, 25 p.). Available from Internet. http://www.ncl.ac.uk/cre/publish/discussionpapers/pdfs/dp28JamborHarvey.pdf.

Jambor, A., & Harvey, D. (2010b). *Review of the challenges of CAP reform* [online] (Centre for Rural Economy Discussion Paper Series No 27). Available from Internet. http://www.ncl.ac.uk/cre/publish/discussionpapers/pdfs/dp27JamborHarvey.pdf.

Jamison, A. (1996). The shaping of the global environmental agenda: The role of non-governmental organizations. In S. Lash, B. Szerszynski, & B. Wynne (Eds.), *Risk, environment, modernity*. London: Sage.

Jamison, A. (2001). *The making of green knowledge. Environmental politics and cultural transformation*. Cambridge University Press.

Jamison, A. (2010). Climate change knowledge and social movement theory. *WIREs Climate Change, 1*, 811–823. Wiley & Sons. 10.1002/wcc.88.

Johnson, S. P., & Corcelle, G. (1989). *The environmental policy of the European Communities*. London: Graham & Trotman.

Káposzta, J., & Nagy, H. (2008). *The perspectives of agriculture in the European Union* [online] (pp. 204–214). Bulletin of the Szent Istvan University. Available from Internet. http://ageconsearch.umn.edu/bitstream/47548/2/24kaposzta-nagy.pdf.

Klein, N. (2000). *No logo: No space, No choice, No Joba* (p. 2000). London: Flamingo.

Knill, C. (2003). *Europäische Umweltpolitik. Steuerungsprobleme und Regulierungsmuster im Mehrebenensys tem*. Opladen: Leske & Budrich.

Koning, N. (2006, November 7–10). Agriculture, development and international trade: Lessons to be learned from the common agricultural policy of the European Union [online]. In *Forum on Food Sovereignty*. Niamey. Available from Internet. http://www.roppa.info/IMG/pdf/Niek_KoningAgriculture_development_and_international_trade-Niamy_November_2006.pdf.

La Via Campesina International Peasant's Movement. https://viacampesina.org/en. 17 October 2020.

LEADER/CLLD. *European network for rural development*. https://enrd.ec.europa.eu/leader-clld_en. 17 October 2020.

Malhi, Y. (2017). The concept of the anthropocene. *Annual Review of Environment and Resources, 42*, 77–104. https://doi.org/10.1146/annurev-environ-102016-060854.

Martinez-Torres, M. E., & Rosset, P. M. (2008). La Via Campesina: transnationalizing peasant struggle and hope. In R. Stahler-Sholk, H. E. Vanden, & G. D. Kuecker (Eds.), *Latin American social movements in the twenty-first century: Resistance, power, and democracy* (pp. 307–322). Lanham, MD: Rowman & Littlefield.

Martinez-Torres, M. E., & Rosset, P. M. (2010). La Vía Campesina: The birth and evolution of a transnational social movement. *The Journal of Peasant Studies, 37*, 149–175. https://doi.org/10.1080/03066150903498804.

Mauss, A. L. (1975). *Social problems as social movements*. Philadelphia: Lippincott.

McMichael, P. (2006). Peasant prospects in the neoliberal age. *New Political Economy, 11*(3), 407–418.

Mertig, A. G., & Dunlap, R. E. (1995). Public approval of environmental protection and other new social movement goals in Western Europe and the United States. *International Journal of Public Opinion Research, 7*(2), 145–156.

Mertig, A. G., & Dunlap, R. E. (2001). Environmentalism, new social movements, and the new class: A cross-national investigation. *Rural Sociology, 66*(1), 113–136.

Mitchell, R. C. (1990). Public opinion and the green lobby. In N. J. Vig & M. E. Kraft (Eds.), *Environmental policy in the 1990s* (pp. 81–99). Washington, DC: CQ Press.

Moehler, R. (2008). The internal and external forces driving CAP reforms. In J. F. M. Sweenen (Ed.), *The perfect storm. The political economy of the Fischler reforms of the common agricultural policy* (pp. 76–82). Brussels: Centre for European Policy Studies.

Patel, R. (2005). Global fascism, revolutionary humanism and the ethics of food sovereignty. *Development, 48*(2), 79–83.

Patel, R. (2006). International agrarian restructuring and the practical ethics of peasant movement solidarity. *Journal of Asian and African Studies, 41*(1/2), 71–93.

Report of the World Commission on Environment and Development Our Common Future. United Nations. (1987). Available from Internet. https://www.are.admin.ch/are/en/home/sustainable-development/international-cooperation/2030agenda/un-_-milestones-in-sustainable-development/1987–brundtland-report.html.

Rootes, C. (1995). Environmental consciousness, institutional structures and political competition in the formation and development of Green parties. In D. Richardson & C. Rootes (eds.), *The green Challenge. The development of green parties in Europe* (pp. 170–184). London and New York: Routledge.

Rosset, P. M., & Martínez-Torres, M. E. (2012). Rural social movements and agroecology: Context, theory, and process. *Ecology and Society, 17*(3), 17. https://doi.org/10.5751/ES-05000-170317.

Sastak Group. https://sastak.com/about-us/what-is-a-machinery-ring/. 17 October 2020.

Schlosberg, D., & Collins, L. B. (2014). From environmental to climate justice: Climate change and the discourse of environmental justice. *WIREs Clim Change. Wiley, 5,* 359–374. https://doi.org/10.1002/wcc.275.

Schrader-Frechette, K. (2002). *Environmental justice: Creating equity, reclaiming democracy.* Cary, NC: Oxford University Press.

Scott, A. (1990). *Ideology and the new social movements.* London: Unwin Hyman.

Slee, B. (2019). *Smart villages and the European green deal: Making the connections*, p. 1. European Network for Rural Development. https://enrd.ec.europa.eu/sites/enrd/files/tg6_smart-villages_sv-green-deal-bill-slee.pdf.

Slow Food International. https://www.slowfood.com/about-us/. 17 October 2020.

Snow, D. A., Rochford, E. B., Worden, S. K., & Benford, R. D. (1986). Frame alignment processes, micromobilization, and movement participation. *American Sociological Review, 51,* 464–481.

Special Eurobarometer 300. (2008). Report. Climate change. Europeans' attitudes towards climate change.

Special Eurobarometer 490. (2019). Report. Climate change.

Takács, I., & Takács-György, K. (2012, October). Cooperation among farmers for cost saving machinery. In: *Agrarian economy and rural development - Realities and perspectives for Romania.* 3rd Edition of the International Symposium, Bucharest, The Research Institute for Agricultural Economy and Rural Development (ICEADR), Bucharest, pp. 328–335. https://www.econstor.eu/bitstream/10419/76817/1/749478039.pdf.

The European Green Deal, COM. (2019). 640 final, Brussels, 11.12.2019.

The European Leader Association for Rural Development (ELARD). (2020). http://elard.eu/.

The European Rural Parliament (ERP). https://europeanruralparliament.com/. 26 October 2020.

URGENCI—Community Supported Agriculture. (2020). https://urgenci.net/. 26 October 2020.

Vidickiene, D., & Melnikiene, R. (2014). *Evolution of rural policy*. Vilnius: Lithuanian Institute of Agrarian Economics.

Vilkė, R., & Šarkutė, L. (2018). Lithuanian local action groups: Spatial initiatives or mobilized potential for rural development? In *Viešoji politika ir administravimas = Public policy and administration*. T. 17, Nr. 4, pp. 510–525. https://doi.org/10.13165/VPA-18-17-4-02.

Ward, N., Atterton, J., Kim, T., Y., Lowe, P., Phillipson, J., & Thompson, N. (2005). Universities, the knowledge economy and 'neo-endogenous rural development' (Centre for Rural Economy Discussion Paper Series No. 1. Endogenous Rural Development). Newcastle: Centre for Rural Economy, Newcastle University. Available from Internet. http://www.ncl.ac.uk/media/wwwnclacuk/centreforruraleconomy/files/discussion-paper-01.pdf.

Wezel, A., Bellon, S., Doré, T., Francis, C., Vallod, D., & David, C. (2009). Agroecology as a science, a movement and a practice. A review. *Agronomy for Sustainable Development, 29*(4), 503–515. Springer Verlag/EDP Sciences/INRA. https://doi.org/10.1051/agro/2009004.

Yúdice, G. (1998). The globalization of culture and the new civil society. In S. Alvarez, E. Dagnino, & A. Escobar (Eds.), *Cultures of politics, politics of cultures: Re-visioning Latin American social movements* (pp. 353–379). Westview: Boulder, CO.

Zobbe, H. (2002, August 28–31). *The economic and historical foundation of the common agricultural policy in Europe* [online]. In Paper of the 10th EAAE Congress "Exploring Diversity in the European Agri-Food System". Zaragoza, Spain. Available from Internet. http://ageconsearch.umn.edu/bitstream/24867/1/cp02mo01.pdf.

CHAPTER 6

Industrial Rural Development Paradigm-Grounded Social Movements

Vitalija Simonaitytė, Rita Vilkė, Erika Ribašauskienė, and Živilė Gedminaitė-Raudonė

6.1 Redistribution of Property

6.1.1 Property Rights and Landowners' Movements Around the World

Property rights are the ability of individuals to accumulate private property, secured by clear laws that are fully enforced by the state. Property rights mean a certain degree to which a country's laws protect private property rights and the degree to which its government enforces those laws (Index of Economic Freedom, 2020). Private property and rights to it are some of the essential aspects of any free-market economy and democratic state. However, there are many cases where state and private property rights can contend and that is one of the reasons why landowners and property owners unite to protect their interests and rights. European Landowners' Organization (ELO) and International Union of Property Owners (UIPI) are great examples of landowners' unions which reveal the goals of such type of organizations par excellence.

International Union of Property Owners was founded in 1923 and it is the leading organization for individual owners and private landlords in Europe. The property owners represented by UIPI, range from individual homeowners to landlords with large property portfolios in the

R. Vilkė et al., *Rural Economic Developments and Social Movements*, Palgrave Advances in Bioeconomy: Economics and Policies,
https://doi.org/10.1007/978-3-030-71983-8_6

private-rented and commercial sectors. UIPI also supports dispossessed property owners in former communist countries. International Union of Property Owners unites five million property owners and 29 national organizations in 28 countries in Europe. The main goal of the UIPI is to increase European policy makers' awareness of the importance of the private, individually owned real estate sector and voice members' interests in new and prospective EU legislation that directly impacts members as well as the whole sector. Climate, environment and energy, construction and technical standards, European economic governance, internal market regulation, property restitution, property taxation and rental market are seven fields of UIPI actions (UIPI).

Created in 1972, the European Landowners' Organization represents a large number of rural family business and enterprises as well as individual actors in Europe involved in activities such as farming and agriculture, forestry and cork, wine production, hunting and fishing as well as water and waste treatment. ELO sets out to encourage sustainable development and management, while promoting biodiversity, sustainable bioenergy, food safety, responsible packaging, and combating climate change. ELO also strives to uphold property rights notably in land use, cities, real estates, historic houses and gardens. European Landowners' Organization unites its members in 29 countries in Europe (ELO).

It is clear that these two organizations are only a few examples of well-established and organized movements of landowners across Europe. However, it is interesting to see, that these two organizations are not only well established, active for a long time, but also their goals show that both organizations are changing, i.e. they are not only representing their members in the field of properties rights, property restitution or taxation, but they are actively involved in advocating for sustainable development, climate change, environment, energy and many other things. It might be one of the reasons why some organizations are still relevant and others after reaching their goal just simply decline.

6.1.2 Case Study 'Lithuanian Landowners' Union'

1st stage: Emergence of social movement 'Lithuanian Landowners' Union'. 'Lithuanian Landowners' Union' was established in 1990. When Lithuania regained its independence in 1990 there have been many issues regarding private property and especially the land. During the Soviet period the land was nationalized and since the 1990s the restitution had to be implemented. The main initiators of the movement were Edvardas

Raugalas, Aloyzas Liaugaudas, prof. Antanas Stancevičius and others, who established 'Lithuanian Landowners' Union' in 1990. At that time, the goal of the social movement was quite modest, but highly relevant—to represent and defend landowners who want to restore the land which was nationalized during the Soviet occupation as of 1940.

Over the years, the activities of the union have been supplemented with new tasks—to promote the acquisition of abandoned lands, to enlarge farms, to represent the interests of Lithuanian landowners in international organizations. 'Lithuanian Landowners' Union' broadened not only the scope of activities but also the movement expanded geographically and had its branches in many districts of the country where people sought for consultations or specific assistance to reclaim their property and other land issues.

In more recent days, as the main goal of social movement has been reached—land rights had been restored, the movement needed more up to date goal. According to the data of the Lithuanian National Land Service, property rights have been restored in rural areas for 782 990 people as of the end of 2018. It is 4.017 mill. ha of land, forest and water bodies. This is 99.82% of the area indicated in citizens' requests. In 46 (out of 60) districts of Lithuania, more than 99.9% of property rights were restored to citizens, except in exceptional cases where property rights could not be restored due to the inactivity of citizens (Ministry of Agriculture, 2018). It proves that the activities of movement had been successful and as the movement reached its main goal, there have been some developments in searching for its new main directions.

The 2010s political agenda and later political decisions revealed another violation of landowners' rights as it was decided to implement Rail Baltica and liquefied natural gas terminal in Lithuanian seaside. It was a good chance for the 'Lithuanian Landowners' Union' to focus on new issues and to protect the interests of landowners in matters of land management, use and disposal in a broader matter. The role of the social movement was seen very clearly in defending landowners' rights in spite of two national constructions which were and are going on in Lithuania: first, it was the building of the main gas pipeline for the liquefied natural gas terminal; second, the construction of Rail Baltica, connecting Warsaw—Tallin through Kaunas and Riga. Implementing the main gas pipeline for the liquefied natural gas terminal, there had been affected almost one hundred of landlords as their land-plot were used for the construction. Many of these landlords felt that their interests were violated. The same situation happened with the construction of Rail Baltica where more than 2500 landlords' rights to their land were affected as the new railway will be constructed on their land (The National Land Service, 2020). So, as the primary goal of the 'Lithuanian Landowners' Union' was reached i.e. rights of the occupied land were restored, the movement experienced that still there is some space for the advocacy of movement's members' rights.

To sum up the first stage of the life cycle—emergence—of social movement 'Lithuanian Landowners' Union' it must be said that the movement implemented its prime goal and managed to maintain its importance by reorienting to new issues and problems in advocating for interests of its members. The main goal of this movement is redistribution of the property when the movement firstly helped members to restore their rights to the land and later on the social movement fought against the uses of rural land for non-agrarian activities (gas pipeline and railway) and fought for the rights of the land owners.

2nd stage: Coalescence and Development of social movement 'Lithuanian Landowners' Union'. The need to gather like-minded people and spread the ideas was a natural further step in reaching the main goal (to represent the rights of union's members in the restoration of the land which was nationalized) as a lot of Lithuanians faced the same problems—i.e. how to restore the land. So, it was a noticeably clear and concrete need for landowners and as many of them faced the same problem it was easier to reach common goals together. That was the main reason why people organized and established a union of landowners.

The number of members has been growing for many years and in 2020 movement had issued about four thousand memberships. However, this number shows the total amount of members and not all members of the movement are active. As the representatives of the organization confirmed—the biggest part of members dropped out after reaching their goal i.e. when a person regains the land usually he or she loses his or her interest in the organization and stops being an active member. However, it shows that the movement 'Lithuanian Landowners' Union' is successful and it implemented its main goals.

The number of active members is much smaller but those who are active, are working vigorously in regions as well as in the largest cities of Lithuania. Also, there can be drawn some common characteristics of 'Lithuanian Landowners' Union' members—the vast majority of them are people living in rural areas, who own land (both small and large farmers), most of them are older than 35 years and mostly men. The higher growth of number of members would be very welcomed but as the reform of land is almost over, the probable future suggests that the number of members of this movement will decrease significantly. Also, there is a substantial change of active members as those who regain the land, usually stop being active members of the movement.

3rd stage: Bureaucratization of social movement 'Lithuanian Landowners' Union. The third cycle stage of any social movement is formalization and bureaucratization where every organization or social movement decides how it organizes its activities, responsibilities and recourses.

Structure of the movement. The movement of 'Lithuanian Landowners' Union' has formal and centralized structure. The union has its own statute, seal, flag, symbolism, and its branches operate in all Lithuanian counties. The highest governing body of the 'Lithuanian Landowners' Union' is the Congress of Representatives, which elects the Council, its chairman, deputies, audit commission and other governing bodies. The movement has its branches and some of them as Vilnius, Biržai, Panevėžys, Šiauliai are more active than the rest of them.

Roles of the movement. Main roles are clearly identified to the members of the movement:

1. To advise landowners on land disposal issues;
2. To prepare proposals to the Parliament and the Government on land restitution, land reform and other agricultural restructuring issues;
3. To represent the interests of its members in all units of land reform administration, as well in the Government, in the Parliament and in the courts of all levels;
4. To organize events and gatherings for the members and for the public (e.g. protests, rallies, etc.);
5. To disseminate information on their activities, land ownership, land reclamation methods and opportunities through all media.

Human resources. Human resources such as labour, experience, employees' skills and expertise are the key elements in reaching the goals of social movement. The social movement of 'Lithuanian Landowners' Union' connected enthusiasts who were active in public life and were willing to solve public issues. For many years the movement had paid staff but as the relevance and the scope of movement decreased, the number of paid staff decreased (in 2015 there had been two employed staff members, in 2016—one) and since the end of 2016, the movement does not have any paid staff.

Financial resources. The sources of funding for the 'Lithuanian Landowners' Union' are enrolment and annual membership fees, as well as funds donated by natural and legal (non-governmental) organizations, interest from credit institutions, and other legally received funds.

Intangible resources. Intangible resources play a major role in the development of any movement. Movement 'Lithuanian Landowners' Union' developed successful informal relationships with many other like-minded movements and other actors which helps them to reach out to stakeholders—landowners and public policy actors and to share their intellectual, human and organizational capital, knowledge and experience in protecting the rights of landowners.

Identification of the movement. Movement 'Lithuanian Landowners' Union' is quite well identified in Lithuania. Members actively spread information about landowners' violated rights, about favourable and unfavourable decisions of courts regarding restoration and nationalization of the land, about meetings and agreements with public policy actors (members of Parliament, Government, ministries, etc.) on various channels of information: social media, newspapers, local and national websites, TV broadcast and radio, meetings and conferences and discussions. Information about 'Lithuanian Landowners' Union' can be found on their Facebook page, also the movement had a webpage but at the moment it is inactive. However, the mentioned communication channels earlier were used more often than they are now. Overall, movement is identified by promoting landowners' rights most.

Communication channels. At the moment, as the webpage of the movement is inactive, Facebook is the main tool used for spreading information about the movement to the wider society. Also, the Facebook page was used for internal communication among members, where they used to find information about future events and meetings. However, at this time the Facebook page is not used actively, as there are not any recent meetings or events organized by the social movement. The more active internal communication between members is going on private bases on phones or emails and only the results of their work are published widely. Movement 'Lithuanian Landowners' Union' does not take part in any international movement at the moment.

4th stage: Decline of social movement 'Lithuanian Landowners' Union'. Finally, the last stage in any social movement's life cycle is decline, which does not necessarily mean failure for social movements. That is why—looking from the scientific perspective—we can see that movement 'Lithuanian Landowners' Union' was highly successful, it had reached its goal and now the movement is declining. The decline of the social movement is met by all the aspects of the movement's activities—the number of members decreased, the numbers of staff decreased to none, the movement does not use any communication channels to connect its members via social media platforms, and finally, the movement does not have its webpage anymore.

There are two possible future scenarios for the movement 'Lithuanian Landowners' Union'. In the first scenario, the future of the movement depends on the political and public agendas—if there are going to arise any new issues regarding the violation of landowners rights—then it is possible to see the revival of the moment. The second scenario of the movement could be related to a broader scope of the understanding of the 'landowners rights'. This could be related to a more recent social movement promoting clean(er) air, water, soil, emphasizing better qualities of neighbourhoods and environments.

This social movement had an impact on the society by:

- representing and advocating the interests of landowners;
- providing consultations on land restitution;
- initiating an appeal to the Ministry of Agriculture and the Government to extend the ban on the sale of land to foreigners;
- assessing the interests of landowners by expressing written opinions and suggestions on the implementation of forest and environmental policy to the Parliament and the Government;
- representing and defending the interests of its members in various courts.

Overall, movement 'Lithuanian Landowners' Union' is a successor of one of the first industrial movements, whose main goal was the redistribution of property and even Lithuanian development was distorted, the movement followed the classical history of industrial movements and it was one of the first social movement established in Lithuania in 1990. The main goal of this movement is redistribution of property and firstly the movement helped members to restore their rights to the land and later on it fought against uses of rural land for non-agrarian activities (gas pipeline and railway). The movement was highly successful, it implemented its goals and it is slowly abiding by the phase of decline.

6.2 Redistribution of Social Power for Social Hierarchy Change

6.2.1 *The Raise of Debate Regarding the Social Hierarchy Change Around the World*

Social movements as drivers of social change are essentially related to social structures with a particular hierarchy which becomes a target of social movement activity (Edelman, 2005; Robles & Veltmeyer, 2015; Rucht, 1996). Particular concerns related to social hierarchy use to be discussed for centuries in a variety of scientific and policy fields, in the context of this research mainly concerning for aspects: the small farmers' movements, rural workers' movements (trade unions), rural women movements and indigenous movements.

Rural workers' movements deserved particular scientific attention in contesting the social hierarchy in Latin America (Petras & Veltmeyer, 2001; Robles, 2001; Robles & Veltmeyer, 2015). In Mexico, Ecuador, Bolivia, Paraguay, Chile and Brazil peasants organized the collective contests of power structures that supports social, political and economic exclusion. By issuing the commitment to guiding objectives of 'land, democracy and social justice' (Robles, 2001), they formed networks with urban trade unions and thus reached the redistribution of social power and social hierarchy. European continent went through another surge of agricultural workers' movements, which is related to decollectivization in Eastern Europe (Mathijs & Swinnen, 1998), as well as free labour force movement after the EU establishment. Agricultural workers' movements raised the concerns regarding the working conditions, workload and wages and contributed to significant changes in the field (Jose, 2013).

Among the most common manifestation of rural women had been related to the roles in the rural husbandry, the wage paid for the work done, compared women and men and many other issues (Hoggart et al., 2014). For instance, African rural women struggled against corporate agribusiness and extractive industries, which aimed to control their resources: land, seed, children and other. At their homes, they also felt pressure and exploitation, as well as in their rural community, in the workplace and even—in the social movements. After successful protests and changes reached in their local communities, African women in 2020 established a huge Rural Woman's Assembly to challenge agribusiness and patriarchal oppression while advancing agroecology and building new leadership for a feminist agroecology on a broader scale (Andrews, 2020).

Scientific studies reveal that gender-related problems regarding the wage in the agricultural industry sector existed for a long time, and still exist in the twenty-first century. Many evidence had been collected that full-time women's hourly wage compared to men varied two times (Hoggart et al., 2014, p. 222). Another issue, which served as a motive for the social movement to arise, was the classical understanding of the roles taken by men and women in rural areas, which also called for changes in social hierarchies.

Recent scientific findings disclose, that small local farmers movements tend to flatten the hierarchical governance structures and, as loosely organized small groups, serve for creating advanced sustainable society 'nests', which affirm local, cultural, environmental, economic and physical way of life (Steup et al., 2018). So, the beginning of the twenty-first century marked a kind of 'rural renaissance', when small agricultural homesteads start being treated again as the most sustainable and resistant way of life and thus deserve exceptional attention in this research.

6.2.2 Case Study 'Lithuanian Family Farmers Union'

1st stage: Emergence of social movement 'Lithuanian family farmers union'. After Lithuania regained independence from the Soviet Union in the 1990s, the number of reforms composed large gaps of farming conditions in the country, which accelerated particular social movements to emerge. Significant country-level motivation factors came after the two most important goals of the agricultural policy after the regained independence:

1. to carry out land reform, to privatize the assets of 'kolkhozes'—collective Soviet farms;
2. to create a new system of agricultural product price and farmers' income support.

In 1991, the Law on Restoration of Citizens' Property Rights to Surviving Real Estate, Land Reform, Agricultural Companies and other actual laws had been adopted in Lithuania. Pursuant to these laws, overall Lithuanian agriculture was reorganized and privatized in two ways:

1. the land and other surviving real estate were returned to the owners or their heirs (restitution);
2. the property created by joint work was privatized by the persons who created it.

A small area of state land was sold to investors. By the end of 2005, the ownership of 797,000 applicants to 3.8 million hectares area was restored; this accounted for 92.5% of the area requested by citizens to restore property rights. Restitution of land and sale of land only to private individuals had an impact on the increase in private land, as legal entities were initially unable to acquire state-owned agricultural land. In 2005, out of 2,747,000 ha of agricultural land, only 1.7% (46,700 ha) belonged to legal entities.

During the privatization of the property of the former 'kolkhozes', a new type of farms was formed during the land reform:

- farmers' farms,
- agricultural companies and other agricultural enterprises,
- homestead family farms (2–3 ha landowners).

In 2005, there were 85,900 farmers' farms, 543 agricultural companies and other enterprises, 153,100 homestead family farms. The assets of former 'kolkhozes' were shredded, divided into technological complexes, and sold to farm workers for general and agricultural investment payments made to them. Farm workers acquired the property mainly in the form of shares.

Inadequate agricultural reform, declining demand and weakening economic ties with the countries of the Commonwealth of Independent States following the collapse of the Soviet Union have led to a sharp decline in agricultural production. The Lithuanian government has taken steps to stabilize it and succeeded.

In 1993, a programme for the development of agricultural (farmers) activities was prepared. It provided for balancing the number of animals and the feed base, providing the population with basic foodstuffs, allocating funds from the state budget for the construction of public roads, electrification and telephony of homesteads, reconstruction of drainage systems, obtaining preferential credits and construction of homesteads. At the 1997 meeting of the Council of Europe in Luxembourg, Lithuanian agriculture became an integral part of the recognized European agricultural model. The most important features of this model—sustainable multifunctional agriculture and rural development, became guiding principles of Lithuanian agricultural policy as well.

After Lithuania became a member of the EU in 2004, agriculture and food development are developed in accordance with the general EU policy. Favourable conditions should be created for stable economic, social and environmental development of agriculture and food, as well as rural development. At the beginning of the twenty-first century, the most important sources of investment in agriculture were the support of the EU Structural Funds, the Rural Development Plan and the Rural Support Program, which aim to facilitate farming in less suitable areas, promote sustainable agriculture and help modernize small farms.

In turn, Lithuania's farms gained the current structure (see Fig. 6.1), which formed particular issues in the light of the agricultural policy proposed by the EU and the Government of the Republic of Lithuania.

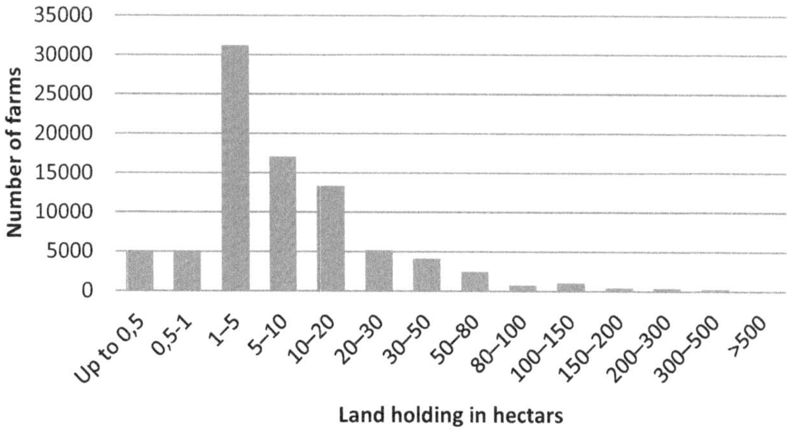

Fig. 6.1 1. Lithuanian farmers' holdings by total area in 2020 (*Source* Agricultural Information and Rural Business Center [2020])

In the year 2020, Lithuania counted 86,000 farmers' farms, who owned 1,337,306 ha agricultural land, or 15.4 ha on average.

Due to three-decade-lasting political, economic and social changes in the country, the structure and conditions of farming greatly varied, thus forming preconditions for the collective behaviour of farmers' groups and consequently—particular issue-related social movements to emerge. The collapse of the previous 'kolkhoz' system and privatization gave birth to huge agricultural companies with concentrated land holdings, agricultural infrastructure and equipment. Others started farming activities on their ancestral land, given back to the rightful hosts after the regained Lithuanian independence. And still, others took up farming activities on newly bought land as a family business.

Year-by-year after the regained independence of Lithuania, small and family farmers started experiencing the increasingly skeptical attitudes towards their activity from the big ones, including land owners and agricultural companies. It was closely related to the ongoing policy reforms regarding the support for agriculture and rural development. The latter increasingly remained apart from the overall country's agricultural policy, which, as experienced by small family farmers, was composed by the government of the Republic of Lithuania, and implemented according to collectively represented big farmers', land owners and agricultural companies' vision and needs, driven by the industrial paradigm. Overall rural development of Lithuania with small family farms and all rural areas-related issues seemingly was moved apart from the key country development policy trends, since the voice of family farms had not been heard by the policy makers.

Therefore the very beginning of the twenty-first century marked for Lithuania the time to unite Lithuanian family farmers and to become a consolidated collective power with the big farmers contradicting voice, which would be heard at the government level when forming the agricultural and rural development policy.

2nd stage: Coalescence and Development of social movement 'Lithuanian family farmers union'. The need to gather like-minded people, to promote and defend the ideas and benefit of family farming was a natural outcome of the overall political, economic and social development in Lithuania, which lasted for a decade since Lithuania regained its independence in 1990. The primary very practical motive for family farmers to unite was to keep a contradicting power against another social movement—Lithuanian Farmers' Union, which started representing only the interests of big farmers and agricultural companies.

The Lithuanian Family Farmers' Union (further—LFFU) was officially launched on February 17, 2000. The LFFU launching congress in the year 2000 was organized in Kaunas, the second biggest city in Lithuania. Among the 170 respective founders of the Union, the well-known Lithuanian family farmers took part in the founding initiative of the movement. Currently, the movement unites nearly 3,500 thousand family farmers throughout Lithuania. The driving force of the movement is composed of some 20 family farmers from different regions of Lithuania, different size family farms and different professions. Among the most active persons are male farmers, whereas women also take part in the activity. Every year some 200 farmers join the union, including the young ones, who newly established their family farms. However, there is an insignificant change in LFFU membership. Most often this is reasoned, as stated by the current chairman of the LFFU Vidas Juodsnukis, with lost belief in the movement's political force, since at the government level, the *LFFU's appeals about the multiple issues regarding the family farms in Lithuania are lastingly ignored.*

3rd stage: Bureaucratization of social movement 'Lithuanian family farmers union'. The third cycle stage of any social movement is institutionalization, formalization and bureaucratization where every organization or social movement decides how it organizes its activities, responsibilities and recourses.

Structure of the movement. Since the establishment in the year 2000, LFFU act as an independent, voluntary public non-profit organization uniting and representing the citizens of the Republic of Lithuania, who have their own farms or agricultural business and most of the agricultural work, except seasonal work, is performed by their family members.

Organizational structure is centralized. The headquarters or LFFU is placed in Kaunas, the same building as Lithuanian Chambers of Agriculture. In the beginning, there was an aim to form regional decentralized representatives of the movement in every district. However, a decentralized system had not proved its durability in turn, since the organization was too resource-consuming as a voluntary activity. The Union has an elected board of representatives, which consist of 31 elected persons from LFFU members. Vidas Juodnukis is currently the chairman of LFFU.

Roles of the movement. The main role of the established LFFU was mobilizing the joint power, which would drive Lithuanian rural areas closer to the EU level by representing the interests of small family farms in rural policy decision-making which directly affect their life. The ambition remains to form sustainable and durable family farms with land holdings from 30 to 40 hectares, which might fully support the needs of the community, the region and contribute to the sustainable and harmonious country's development, as well as create peaceful conditions for family to work and live with dignity.

More precisely, the LFFU role is disclosed in seven objectives:

1. To represent and defend the interests of members.
2. To take care of the professional development, professional development and general farming culture of farmers, especially their younger family members.
3. To ensure that national and European support for the development of family farms is provided on favourable terms.
4. To strive for the Chamber of Agriculture, advisory services, to provide advice to family farms of a scientific, economic, legal, professional and educational nature.
5. To fully support the self-government, citizenship, cooperative movement and spiritual rebirth of emerging farmers and all rural people.
6. To submit proposals to the Seimas (the Parliament) of the Republic of Lithuania, the Government of the Republic of Lithuania and other institutions regarding the development of family farms, improvement of laws and by-laws.
7. To communicate and share good practice with foreign organizations, institutions, firms of similar profile. To actively participate in the activities of the European landowners—ELO organization.
8. To organize farmers' markets, where farmers could sell their products.

The initial goal of the 'Lithuanian family farmers union' well corresponded to the overall activity of the movement and had not been changed since the establishment.

Human resources. The LFFU unites Lithuanian family farmers with different skills, experiences, education and professional capabilities. The governing structure is formed from internal movement human resources—elected ambitious and skilled family farmers. The new family farmers, who join the movement, have an ambition to improve their farming skills, to share the ideas, products and practices within the movement as well as to get involved in political solving of their activity-related issues and in general family farming in Lithuania. Participants of the movement strive to improve their skills and competencies via possible open courses, seminars. As stated by LFFU chairman Vidas Juodsnukis, a huge demand for family farmers' consultations regarding small and family agribusiness is observed. The existing farmers' consultation system in Lithuania in the forefront by huge public enterprise 'Lithuanian Agricultural Advisory Service' is concerned with industrial farmers' issues, whereas small family farmers feel left behind.

Financial resources. LFFU is an open public organization, which does not hold any right to targeted direct state financing. Financial resources of the movement consist of voluntary collected membership fee, which is symbolic and annually equals 10 euros. Some movement members pay the fee annually, some miss it. Any additional expenses of the movement for particular events and other activities are organized at that time by activity initiators.

Intangible resources. LFFU quite well developed the informal relationships, which highly helped promoting the short supply chain and direct sales via so-called 'Farmers' markets'. This activity became very popular in urban areas, especially in Lithuanian biggest cities. Normally farmers' markets visit different locations in the city next to the big shopping centres on the scheduled days. Thus farmers developed a huge network of consumers, who value fresh and healthy food directly from the farm. Thus the LFFU via farmers' markets promotes their main movement's ideas, concerning the added value and benefits of family farms: products, grown and produced in family farms are made in a sustainable and environmentally friendly way.

Identification of the movement. LFFU is wel-identified in Lithuania at the policy level, since they constantly propose many ideas regarding the key agricultural and rural development issues in Lithuania and a variety of solutions in the field to different governance structures of the state, including the Seimas of the Republic of Lithuania, The Agricultural Ministry, The Committee of Rural affairs, The Chambers of Agriculture, etc. Regionally, LFFU is also identified quite well, since annually they organize a round of sessions throughout the whole Lithuania and discuss there the news regarding the family farming policy and practices and main issues in regions. Indeed, the general citizens better know and use the successful initiative of LFFU farmers' markets, spread throughout the whole Lithuania rather than the LFFU itself. News about the LFFU and benefits of family farming from time to time appears in TV and radio interviews.

Communication channels. LFFU uses particular internal and external communication channels. Information about LFFU can be found on two of their websites (www.semosukiai.lt, www.ukiai.lt), as well as in the Facebook group 'Lithuanian Family Farmers Union'. The websites provide the list of Lithuanian family farmers and propose the possibility to contact them regarding the fresh production directly. The Facebook group of LFFU is quite well-identified and quite actively exploited by Lithuanian citizens as a place for spreading the ideas of sustainable family farming as a future of human beings, as well as a place where consumers can find information regarding the direct sales of fresh and healthy food directly from the farmer. Currently, the Facebook group platform is very actively used. It has more than 9,300 thousand followers of the posted news.

International activities. As stated among the main roles of LFFU, movement members use to actively participate in the activities of the European landowners—ELO organization for a long period of time. This international organization helped the LFFU get acquainted with the newest information regarding the family farming in the EU, helped in improving qualifications and skills, sharing the hot issues regarding the family farming in Lithuania and the broader. LFFU felt a great inspiration from the organization to keep promoting their ideas of sustainable family farming and related issues. However, a few years ago the membership was charged with high fee under the pressure of the biggest landowners and agribusiness of the EU, and LFFU was forced to stop their membership in the organization. This recently make a negative effect on the overall activity of LFFU, especially on the soul and identity of the movement, which suffers again from ignorance, especially from the government side.

4th stage: Decline of social movement 'Lithuanian family farmers union'. Current stage of the development of the LFFU social movement might be evaluated from the two aspects. Part of the LFFU social movement, which aimed to connect family farmers with direct consumers and spread the idea of sustainable living with fresh and healthy food, became a mainstream due to the amounts connected people via direct sales in a form of Famers' markets throughout the whole country. This definitely signalizes the *success* of LFFU in the field. From the other, i.e. impact's to family farming policy in the context of the whole EU, the movement performs failure current times. International activity had been ceased. As stated by the current chairman of the movement Vidas Juodsnukis, especially the latter times when the big agribusiness representatives took political power in the Seimas and Government of the Republic of Lithuania, LFFU went through really hard times and oppression. Indeed, the period of the year 2004–2008, when the Agricultural policy was guided by professor Kazimira Prunskienė, might be considered the golden age of the movement, since the voice of LFFU then was best heard and treated by all Lithuanian government.

6.3 Redistribution of Added Value Between Farmers and Other Agents of Supply Chain

6.3.1 Associations of Agribusinesses in Europe

Agribusinesses and agricultural companies play a major role in European agricultural policy and there is a huge variety of unions and associations of agribusinesses. The biggest part of such associations unites sectorial agricultural companies, such as crops, seeds, meat production and processing and many others. Another major role is played by national associations of agricultural companies and such organizations are established in each and every European country. CELCAA, the European Liaison Committee for Agricultural and Agri-Food Trade and 'Euroseeds' are good examples of such associations acting on the European level and Agricultural Industries Confederation (AIC) represents an example of national case, as it serves the agribusiness and agricultural supply industries across the United Kingdom.

CELCAA, the European Liaison Committee for Agricultural and Agri-Food Trade is the umbrella organization representing at European level associations and companies active in the sector of agricultural and agri-food trading. Its full and affiliated members include cooperative and non-cooperative wholesale traders (collectors, distributors, warehouse keeper, importers and exporters) delivering agricultural and agri-food products as feed materials to farmers and compound feed industry, as well as raw material to the food industry, as food and drink to retailers. CELCAA promotes the interests of the European agricultural and agri-food traders at the European level towards EU institutions as well as towards other associations representing other sectors of the food supply chain. CELCAA provides members with the latest legislative and technical developments on various policy issues common to agri-produce trade. The organization also works as a networking platform for its members to exchange information and expertise and helps to identify the right interlocutors in the sector and in European Institutions. CELCAA membership is twofold, with full members and affiliated members. All trade associations in agri-food and agri-commodities being European or national can become a member of CELCAA (CELCAA).

Euroseeds is the association of the European seed sector. The organization represents the interests of those active in research, breeding, production and marketing of seeds of agricultural, horticultural and ornamental plant species. Euroseeds addresses a number of key topics, from food security, to food quality and health; from jobs and growth to sustainability and environmental protection. Euroseeds, with more than 34 national member associations from EU Members States and beyond, represents several thousand seed businesses, as well as 67 direct company members, including from seed-related industries (Euroseeds).

The Agricultural Industries Confederation (AIC) is the agri supply industry's leading trade association in the United Kingdom. Formed in 2003 by a merger of three trade associations, the trade association represents several sectors within the agri supply industry, including animal feed, crop protection and agronomy, fertilizers, grain and oilseed, seed. AIC works on behalf of its members by lobbying policy makers and stakeholders, delivering information, providing trade assurance and offering technical support (AIC).

6.3.2 *Case Study 'Lithuanian Association of Agricultural Companies'*

1st stage: Emergence of social movement 'Lithuanian Association of Agricultural Companies'. After the restoration of independence in Lithuania, the agricultural structure has changed significantly as individual farms and private agricultural companies emerged. The government was expected to facilitate conditions for farmers to become real landowners, enable them to farm with dignity in all possible forms: individually or collectively—through agricultural enterprises (cooperatives). But that did not happen as these expectations were largely ignored. Thus the idea to unite in the Association was born, in order to promote the interests of farmers and agricultural enterprises.

The initiative, which stemmed from the agricultural companies themselves, aimed to ensure the representation of their interests at all levels of government.

The most important goal of the Association has been to strive for an enabling economic, legal, financial and social environment for farmers and for favourable market conditions for agricultural companies. The main goal of the Association of Agricultural Companies has remained the same since the establishment of the Association.

In order to successfully implement its key goal, the Lithuanian Association of Agricultural Companies has envisaged these main types of activities: to express and promote the interests of its members and to represent them in state institutions and non-governmental organizations; to improve the activities of agricultural companies and other companies working in a similar field; to raise the qualification of farm managers and specialists who are the members of the Association; to look for partners and funds for the implementation of investment projects, dissemination of new equipment and technology; to develop innovative programmes for growing and selling agricultural products to Lithuanian and foreign partners; to improve relations with the committees of Seimas of the Republic of Lithuania, the Government of the Republic of Lithuania, Ministries, various organizations representing agricultural and food-processing industry, as well as foreign companies and their representative offices; to inform farmers and rural residents of the requirements of the European Union; to represent Lithuanian agricultural companies in Lithuania and abroad, and to facilitate collaboration with similar organizations in Lithuania and other countries.

In 1992, the Lithuanian Association of Agricultural Companies was established by agricultural companies and the scientific community. The Association has quickly become recognized in Lithuania and abroad as representative body of agricultural companies.

2nd stage: Coalescence of social movement 'Lithuanian Association of Agricultural Companies'. The Association of Lithuanian Agricultural Companies has united like-minded people and become a force. Jonas Sviderskis, the Director-General of the Association, stated that if it were not for establishing this Association back in 1992, it would be difficult to estimate how many agricultural companies producing commercial agricultural products would be today. The decision to unite was very important, timely and significant not only for the agricultural companies operating at that time, but also for the whole agriculture of the country.

Participation in the activities of the Association have provided a greater opportunity to reconcile the goals and actions of a large group of individual farmers and agricultural enterprises; to represent the interests of farmers more effectively; to increase cost-saving; to connect with authorities and shape the relevant legislation. The fact that farmers in other countries of the European Union have been actively involved in self-government has also stimulated the formation of a circle of like-minded people. In addition, membership in the Association makes it easier for members to engage with policymakers and other businesses.

Membership in the Lithuanian Association of Agricultural Companies is open. Currently, the Association unites 159 agricultural, private companies, cooperatives and emerging agricultural companies. 35 farmers also participate in the activities of the Association. In total, the Association has 215 members. The vast majority of the members of the Association are active.

In addition to agricultural companies, farmers and enterprises providing agricultural services also participate in the activities of the Lithuanian Association of Agricultural Companies, as well as farms engaged in the activities of breeding pigs and horses, orchards and fruit production, farms belonging to scientific institutions, etc. Agricultural companies which are the members of the Association work only 13% of the land, but are highly productive and produce about 35% of commercial, high-quality agricultural products. The average number of shareholders in the member company is 16 persons. Members of the Lithuanian Association of Agricultural Companies have hired over 17,300 employees who are socially insured and receive all the social guarantees to which they are entitled. On average, 34 people work on one farm. The average employee age is 49 years old, and workers aged up to 40 accounts for about 30% of all workers. The age structure of employees varies greatly from company to company. The turnover of employees is low and the average length of service is 23 years. On average, the company employs about 34 people and has 28 ha of agricultural land per employee. 94% of managers have acquired agricultural education (96% of them have higher education, 4% of them have technical education) and 92% of the leading technical personnel have acquired agricultural education (of whom higher education—30%, technical education—65%).

The Lithuanian Association of Agricultural Companies is a growing organization. Since 2000, 1–2 new members join the Association each year. Large-scale farmers, companies providing agricultural services and other associations engaged in agricultural activities have also joined the activities of the Association. The Association is becoming more diverse in terms of membership. It can be said that the membership in the Lithuanian Association of Agricultural Companies promotes successful performance goals, increases prestige and bargaining power. The members of the Presidium of the Association are well-known farmers in Lithuania.

3rd stage. Biureaucratization of social movement 'Lithuanian Association of Agricultural Companies'. *Resources of the movement.* Jonas Sviderskis, Director-General of the Lithuanian Association of Agricultural Companies, noted that the cost of membership in the Association is not high and the budget of the Association is relatively stable. The Association has its headquarters in Vilnius, and hires employees to support its work.

The Presidium of the Association is formed on the territorial and sectoral basis, with the aim to represent the entire country. The Association has been mobilizing the necessary resources successfully, as each member of the Presidium focuses primarily on resource mobilization activities in their respective region and in addition, aligns these activities to their respective area of expertise.

Structure of the movement. The Lithuanian Association of Agricultural Companies is a voluntary and non-governmental non-profit organization, operating under the Lithuanian law of associations and other legal acts. Decisions taken by the Congress of members are implemented and activities are organized by the seventeen members of the Presidium of the Association, under the leadership of the President of the Association and two Vice Presidents.

Membership. Membership in the Lithuanian Association of Agricultural Companies is open. An invitation to become a member of the Association of Lithuanian Agricultural Companies has been posted on the Association's website, and the application can be submitted by filling in the electronic membership form and sending it by e-mail or delivering it to the Association's headquarters. The requests for the membership are considered at the meetings of the Presidium. The Association aims that the membership would not be driven only by short-term considerations to acquire benefits, but also would help to achieve the unity and solidarity of members to achieve common goals.

Roles of the movement. The Association hires employees to undertake everyday activities. The Association has a General-Director, Deputy Director-General, Financial Officer and Referent. Director-General and the Deputy represent the Association at the Working Groups of the Ministry of Agriculture, including Working Group for biological assets and for setting up normative prices for agricultural produce; Project Selection Committee, where projects are considered in the following areas: support for investment in agricultural holdings; support for investment in the processing, marketing and/or production of agricultural products; support for the creation and development of European Innovation Partnership (EIP) action groups; at the State Council for Agriculture and Research; at the Committee monitoring the implementation of the Lithuanian Rural Development Program 2014–2020, etc.

Identification of the movement. In addition to the annual conventions, the Association organizes weekly conferences for its members every year. Seminars are held separately according to the need of the members. Sometimes regional events are held too. Usually, information on financial, tax, health and safety, environmental protection matters is shared with members. Traditional events are organized to mark the anniversary of the Association. In addition to the members of the Association, high-level government officials, social partners and foreign partners are invited to participate. The website of the Association provides information on the events that took place, key decisions of the Association and other relevant information.

Communication channels. Daily communication is by electronic mail and telephone. Extended meetings of the Presidium are also organized to ensure internal communication. The Association is a permanent member of many working groups in the Government of Lithuania, Seimas and Ministries. The Association regularly participates in the meetings of the Committees of Seimas, including the Committee on Rural Affairs, Environmental Protection and other committees, many governmental meetings, inter-institutional meetings. All the latest information or decisions by Association having an impact on its members are uploaded to the Association's website—sections *News* and *For Registered Users*. The information is constantly updated.

The Association has produced several publications for managers and specialists: 'A set of documents regulating the activities of agricultural companies', 'Calculation of production costs', 'Overview of the activities of agricultural enterprises and other companies', 'Compendium of legal acts and explanations for legal entities when acquiring land' and others.

Scientists, politicians and government representatives are invited to the meetings of the Presidium of the Association. Usually, major events are attended by senior government officials, including the President, the Prime Minister or the Speaker of Seimas.

The Association does not have any active virtual platforms.

Participation in the international movement. The Lithuanian Association of Agricultural Companies actively participates in the activities of the COPA-COGECA working and advisory groups on Common Agricultural Policy; on Milk and Milk Products; International Aspects of Agriculture; Environmental Protection; Science and Research. It is also a permanent member of many working groups at the European Commission.

4th stage: Decline of social movement 'Lithuanian Association of Agricultural Companies'. According to Director-General Jonas Sviderskis, the activities of the Lithuanian Association of Agricultural Companies are successful and expanding. The Lithuanian Association of Agricultural Companies monitors and evaluates the necessary changes in the Association itself, the opportunities for joint activities and the influx of new members.

The activities of the Association influence policy decisions which may have an impact on the activities of farmers. The Association also always advocates for innovative solutions, the use of environment-friendly farming technologies, and sustainable management of natural resources. The Association guides its members and promotes the application of more advanced methods and technologies and the use of scientific knowledge in farming. As a result, agricultural companies, members of the Association, tend to pursue more diversified agricultural production, and preserve and develop animal husbandry, thus ensuring permanent jobs and incomes, and also contributing to land improvement in Lithuania. The activities of the Association of Agricultural Companies are not limited to the promotion of agricultural production and representation of the interests of agricultural companies, as the Association also takes care of various rural affairs and fostering of human resources.

6.4 Redistribution of Risk Between Farming as Livelihood and Other Sectors

6.4.1 *International and Regional Farmers' Unions and Associations*

There are a huge amount and variety of farmers unions and associations, where each country has at least a few farmers unions, as well as there are regional and worldwide active associations. Many regional and international unions act as umbrella organizations having members of national farmers organizations. Such examples are World Farmers' Organization (WFO) and Committee of Professional Agricultural Organisations—General Committee for Agricultural Cooperation in the European Union (COPA-COGECA).

World Farmers' Organization was established in 2011 and it is a member-based association, bringing together national farmers' organizations and agricultural cooperatives from all over the world. WFO mission is to represent the farmers' voice and advocate on their behalf in all the relevant international processes affecting their present and their future, ranging from the global dialogue on agriculture to nutrition and sustainability. This includes Climate Change negotiations (COP), Committee on World Food Security (CFS), United Nations Disaster Risk Reduction (Sendai Framework), IFAD's farmers' forum and many more. Organization'sThe organization's objective is to create the conditions for the adoption of policies and programmes that can improve the economic environment and livelihood of producers and rural communities, strengthening the contribution of agriculture in tackling the challenges humankind faces (WFO). World Farmers' Organization unites members from 73 countries from six regions: Europe, Asia, Oceania, Africa, Latin America and North America.

COPA-COGECA is exceptionally focused on European farmers organization and unites two organizations: Committee of Professional Agricultural Organisations (COPA) and General Committee for Agricultural Cooperation in the European Union (COGECA). In 1958, the first European representative organization, COPA, was created and one year later, in 1959, the agricultural cooperatives of the European Community created their European umbrella organization, COGECA. COPA unites 60 organizations from the countries of the European Union and 36 partner organizations from other European countries such as Iceland, Norway, Switzerland and Turkey. This broad membership allows COPA to represent both the general and specific interests of farmers in the European Union. Since its inception, COPA has been recognized by the Community authorities as the organization speaking on behalf of the European agricultural sector as a whole. There are four objectives of COPA: to examine any matters related to the development of the Common Agricultural Policy; to represent the interests of the agricultural sector as a whole; to seek solutions, which are of common interest; and to maintain and develop relations with the Community authorities and with any other representative organizations or social partners established at European level. When COGECA was created, it was made up of six members now has 35 full members and four affiliated members from the EU. COGECA also has 36 partner members. In line with the recent European Union enlargements, COPA and COGECA have together further reinforced their position as Europe's strongest farming representative organisations. COPA and COGECA have jointly welcomed 38 national farmer and cooperative organizations from the new Member States. Overall membership of both organizations has thus risen to 76 organizations from the EU Member States (COPA-COGECA). Lithuanian farmers' union and the Lithuanian Association of Agricultural Companies are members of the COPA-COGECA organization and participate in the activities of the organization.

6.4.2 *Case Study 'Lithuanian Farmers' Union'*

1st stage: Emergence of social movement 'Lithuanian farmers' Union'. The Lithuanian Farmers' Union (LFU) is the oldest farmers' organization in the country and it is also the first restored farmers' organization after the restoration of Lithuania's Independence.

LFU was founded in 1919, and its statutes were registered on December 20 of the same year. The first chairman of the organization was Aleksandras Stulginskis, who was elected as the President of the Republic of Lithuania in 1922. After Aleksandras Stulginskis, LFU was led by one of the founders of the Union, Eliziejus Draugelis, who was a doctor and a prominent figure in the Lithuanian state and society. In 1919, the organization had five district branches, in 1925—380, and in 1927—124 district branches. At that time, LFU also established a number of cooperatives, dairy, fruit and flax processing, animal husbandry and grain processing enterprises, warehouses for the storage of agricultural products, and operated small banks. The economic organizations established by the Union operated at its disposal, including the Central Farmers 'Bank (1924–1930), Lithuanian Cooperative Centre (1923–1930), Dairy Union (1926–1928), Lithuanian Farmers' Cooperative Union. In June 1940 after the Soviet Union occupied Lithuania, the Union was dissolved and many of its leaders were imprisoned or deported. After the restoration of Lithuania's Independence, LFU was restored.

In 1988, farmers' movement started in rural areas of Lithuania and LFU was registered as a public organization in charge of coordinating activities of farmers. Its members could become farmers or persons interested in farming, also farm employees. During the first Congress of the restored LFU, 200 people became members of the organization. In 1990, after the entry into force of the Law on Farms, 7 000 members joined LFU. In 1990, LFU was established in all districts and had 44 district branches.

The restored LFU initiated the development of the Law on Farms, contributed to the release of the newspaper 'Farmer's Adviser', established credit unions in Lithuania, and was a shareholder of the newspaper 'Farmer's Newspaper'. LFU also became a founder and shareholder of the Lithuanian Agricultural Advisory Service. Lithuanian farmers belonging to the Union sought to help other farmers to understand the changing environment, to represent farmers' interests in shaping the country's agricultural policy, promote the engagement in agriculture and inform on the policies of the European Union, also to assist farmers in buying the second-hand advanced farming technologies from the Nordic countries which were vital to restoring farms. The Union operated through its branches in each district of the country, which united farmers and those interested in farming. After the restoration of Lithuania's Independence, and with new possibilities to restore land ownership rights, many people started farming. Many of these new farmers did not have a proper agricultural education; they lacked the knowledge of how to profitably develop the farm. In 1992 LFU started a close cooperation with the Federation of Swedish Farmers (LFR), and in 1997, with the Swedish International Development Cooperation Agency (SIDA). LRF training department jointly with LFU implemented a variety of training and educational projects targeting Lithuanian farmers. Main projects included Management School; Strengthening the Women's Network of LFU; Strengthening LFU; Project Management; Development of the Information Network of LFU; trainings for the leaders of LFU.

The main objective of LFU has always been to represent farmers' interests in shaping the country's agricultural policy. At present, LFU is an independent and voluntary Union uniting associations of Lithuanian farmers and persons related to farming sector and interested in farming. Its goals are to coordinate the activities of the Union members, and to represent and promote the interests of the members in various Lithuanian and international institutions. LFU is an important social partner of the

Ministry of Agriculture. On February 12, 1989 the restorative congress of LFU was convened and the statute of LFU was registered on September 17, 1991.

2nd stage: Coalescence of social movement 'Lithuanian farmers' Union'. LFU was established in 1919 and restored in 1989 and on both occasions, Lithuanian agricultural academic community was largely involved. Founders of the Union had a broad vision and understanding of the importance of having an organization, which would represent farmers, would define agriculture as a specific area of activity and help to develop it while adapting advanced agricultural technologies and methods, and promoting scientific knowledge in farming. There was quite a lot of uncertainty at the time of the restoration of Independence, as the collective farm system that had existed for decades collapsed, and there were many questions about what to do with the returned land, what to cultivate, which farming technologies to use. The new generation of today's farmers still remembers those days when farming was extremely difficult. Furthermore, the political instability at that time also brought frequent changes of legislation and introduction of new taxes. Thanks to close collaboration of its members, LFU was able to share ideas and best practices; provide opportunities for farmers to express their interests and to work together to achieve more favourable farming conditions.

As of October 1, 2020, LFU has 41 members, which are legal entities operating in various districts of Lithuania. Farmers participate in LFU through membership in regional farmers' unions. It is estimated that they bring together more than 5,000 farmers.

As of October 1, 2020, the members of LFU are of diverse age and specialization, and manage farms of varied sizes. The areas of land cultivated by members range from 1.5 ha to 1000 ha. Members are producers of commercial agricultural products, who derive their income mainly from agricultural production. Some farmers are also members of cooperatives.

The membership both in LFU and in the regional farmers' unions is conditioned by members' compliance with LFU's obligations. According to the statutes of the Union, a member may withdraw from the organization at any time and should resign if he does not carry out membership obligations properly or without a valid reason, does not pay the membership fee for two consecutive years.

3rd stage: Biureaucratization of social movement 'Lithuanian farmers' Union'. *Resources of the movement.* LFU has its headquarters in Kaunas, where it rents premises. Employees are hired to organize daily activities of the Union. The main income of the Union is the membership fees collected from members, which are approved annually at the Congress and depend on the area of agricultural land cultivated by the farmers of the district. There are also project activities, but project implementation is not the main function of the organization.

LFU has the status of a beneficiary (i.e. Lithuanian residents can allocate a part of the declared personal income tax to the Union, while business enterprises support specific events organized by LFU). The Chairman and Vice–chairmen of the Union work without remuneration.

Structure of the movement. The structure of LFU allows performing its functions as efficiently, flexibly and extensively as possible. The functions are defined in the statutes of the Union. The bodies of the Union are the Congress, the Presidium and the Chairman. The governing bodies are the Presidium and the Chairman. Union's financial activities are managed by the auditor. By the decision of Union's Congress or the Presidium or Chairman of LFU ad hoc (special) groups may be established to consider concrete issues or perform some of the management body's functions.

Membership. Members of LFU may be associations of Lithuanian farmers and persons related to farming and interested in farming who have voluntarily joined the Union and pay the membership fee in a proper and timely manner. Members shall be admitted by the decision of the Presidium upon submitting a written request. After joining the Union, the applying entity pays the admission fee and gets a certificate. The entity becomes a member of the Union from the date specified in the certificate. Individual persons wishing to become members of LFU must apply to the farmers' union of their district. The procedure for their acceptance and resignation is regulated by the respective statutes of the district farmers' union. The Union's website contains all relevant information on the arrangements for admitting members, admission fees and membership fees.

Roles of the movement. In the regions, LFU operates through its members—regional farmers' unions, led by the chairmen. At the national level, LFU operates through the persons authorized by the Congress—the Chairman and Vice-Chairmen.

LFU is engaged in the international cooperation and sends their representatives or collaborators from other agricultural organizations as experts to the relevant COPA-COGECA working group meetings depending on the issues under discussion. LFU's representatives are members of the COPA-COGECA working groups and also participate in the European Commission's advisory groups, also join seminars, conferences and other events where they present the views of Lithuanian farmers and promote their interests, discuss agricultural market conditions and use opportunities for Lithuanian farmers to play an important role in shaping the Common Agricultural Policy of the European Union. The main goal of these LFU experts is to gain as much as possible experience and useful information, and to develop a network of communication and cooperation with other countries of the European Union, which would allow exchanging information on the relevant developments of the agricultural sector. Experts disseminate the received information among Lithuanian farmers through their organizations. Based on the information received, they also liaise with the Lithuanian authorities in order to ensure decent working conditions for farmers, advise on setting appropriate sale prices for products and services, creating conditions for stable income generation and providing relevant information on agricultural developments.

LFU's daily activities are operated by hired administrative staff: Accountant-Administrator, International Relations Coordinator, Corporate Affairs Specialist and Chairman who works without remuneration.

Identification of the movement. In addition to the annual Congress, the Union organizes Presidium meetings, which also usually involve legal experts from regional farmers' unions. These meetings provide opportunities to discuss various issues and listen to diverse opinions. The activities of the Union are geographically localized and regional farmers' unions organize annual meetings for their members—farmers. Decisions taken at these meetings are published on the website.

Since 1994, the traditional 'Farm of the Year' competition has been organized and the winners of last year's competition are visited. Members of LFU are active participants in agricultural exhibitions and events in Lithuania and abroad. LFU also traditionally organizes events to mark the anniversary of the Union. In addition to the members of the Union, the prominent government officials of the country, social partners, and foreign partners, mayors of district municipalities and heads of agricultural departments, heads and representatives of various organizations and associations, sponsors are invited and participate. Information on past events is posted on LFU's website.

Communication channels. Daily communication is done by e-mail and telephone. The Union is a permanent member of many working groups in the government of the country, the Seimas and ministries. Its representatives participate in the meetings of the Ministry of Agriculture, Committees on Rural Affairs, Environmental Protection, Budget and Finance at the Seimas, various Government meetings, and inter-institutional meetings. All information relevant to members is posted on the Union's website sections *News* and *Actualities*, and on LFU's Facebook page. The information is constantly updated. The Union invites scholars, politicians, government officials, agricultural company representatives to the meetings of the Presidium.

Participation in the international movement. LFU became interested in the membership in COPA-COGECA, one of the most influential lobbyist groups in the European Union, during the period of Lithuania's preparation to join the European Union. Invited by Western countries in 2002, LFU became an associate member of the COPA-COGECA organization, gaining an equal footing with the other European Union Member States to represent the interests of Lithuanian farmers in the European Union. Through participation in COPA-COCECA activities, LFU has opportunities to provide comments on the European Union legislation, promote interests of Lithuanian farmers, raise problematic questions on complex agricultural sector problems and provide input to their solutions, develop common positions and thus to play an active role in shaping the Common Agricultural Policy.

LFU became a full member of GEOPA-COPA in 2004. Since then the Union has been participating in the European Commission's Agricultural Social Dialogue Committee meetings, organizes joint seminars and conferences with social partners. Relevant issues discussed at the events help to create a positive working culture in the Lithuanian agricultural sector, to keep pace with other European Union Member States to develop and implement stable, fair and productive labour relations in the sector. The strengthened and modernized Lithuanian agriculture is becoming a strong economic engine of the country. The farmer-employer is perceived as an important social partner, creating jobs and contributing to the development of rural areas.

LFU cooperates with all farmers' organizations of European countries—members of COPA-COGECA. LFU is a founding member of the Council of Baltic Farmers' Organizations. Due to many common interests, LFU maintains the closest contacts with the Estonian and Latvian farmers' organizations.

4th stage: Decline of social movement 'Lithuanian farmers' Union'. There is some competition in Lithuania between numerous organizations representing different agriculture sectors and farmers' interests and associations aiming to unite them. However, LFU remains one of the most influential farmers' organizations operating in Lithuania. LFU engages in participatory and democratic approaches to consolidate the opinions of farmer's managing very different types of farms, and communicates their views to the country's governing institutions, society and the institutions of the European Union.

The views of LFU members do not always coincide with those of other agricultural organizations and are not always taken into account by politicians. Sometimes decisions made by politicians have long-term negative impact on the agricultural sector. Many LFU members are critical towards the public policy, and call for a long-term agricultural development strategy, evidence-based policies, and thorough consultations before introducing tax changes or planning their increases. Overall, Lithuanian institutions responsible for the preparation of the National Strategic Plan for Lithuanian Agriculture and Rural Development for 2021–2027 should pay more consideration to farmers and their contribution to the country's budget, also consider farmers' potential to contribute to solving various economic and social issues in rural areas. The Common Agricultural Policy should retain and strengthen its focus on farmer, and the sustainable development of farmer's activities, thus ensuring conditions to generate incomes from farming, regardless of the specialization of farms.

6.5 Redistribution of Economic Power by Collective Actions

6.5.1 *Movement of Agricultural Cooperatives*

Agricultural cooperatives with the main role to redistribute economic power by collective actions remain significant actors for many decades not only for the EU agriculture but also globally. The cooperative movement has been particularly active in rural areas, as new incentives for the establishment or enlargement of farmers' cooperatives have emerged since a cer-tain degree of agricultural industrialization was reached. In particular, small farmers were encouraged to cooperate with each other as a response to changes that occurred in the market (Vidickiene et al., 2016). Oligopoly or monopsony, with a large number of small farmers but only a few or one agricultural product collector and processor, more frequently appeared in the market of agricultural products as a result of expanded food-processing companies. Oligopoly or monopsony also appeared more frequently in the labour market in some rural regions since the mechanization of agricultural production processes resulted in less employers being willing to employ agri-cultural workers (Vidickiene et al., 2016). The number of cooperatives and scale of movement of agricultural cooperatives differ in old and new EU countries because of the experience of forced cooperation during Soviet times but intentions for cooperation are high.

The movement of agricultural cooperatives have started their first initiatives for networking when the framework for Common Agricultural Policy (the CAP) was created in the Treaty of Rome in 1957. Agricultural organizations from different EU countries were invited by the EU Commission to attend various events on topics related to agricultural policy as observers aiming to initiate close cooperation. As a result, the first European organization representing farmers, COPA (Committee of Professional Agricultural Organisations), was created on 6 September 1958 (Copa-Cogeca, 2020). Next year, on 24 September 1959, the national agricultural cooperative organizations created their European umbrella organization—COGECA (General Committee for Agricultural Cooperation in the European Union)—which also includes fisheries cooperatives (Copa-Cogeca (2), 2020). Six members have created COGECA. Before the EU's enlargement in 2004, it has been enlarged by almost six times and has 35 full members and 4 affiliated members from the EU. After the EU's enlargement in 2004, the COPA and COGECA have jointly welcomed 38 national farmer and cooperative organizations from the new Member States. Overall membership of both organisations has thus risen to 76 organizations from the EU Member States. COGECA, now called the 'General Confederation of Agricultural Cooperatives in the European Union', currently represents the general and specific interests of some 40,000 farmers' cooperatives employing some 660,000 people and with a global annual turnover in excess of three hundred billion euros throughout the enlarged Europe (Copa-Cogeca, 2020). Since its creation, COGECA has been recognized by the European Institutions as the main representative body and indeed the spokesman for the entire agricultural and fisheries cooperative sector. Currently, COPA and COGECA have together further reinforced their position as Europe's strongest farming representative organizations (Table 6.1).

Table 6.1 1 Role of agricultural cooperatives by COPA and COGECA

Cooperatives' activities are founded on the principles of economic democracy, transparency and solidarity among themselves and with their local rural community.
Agricultural cooperatives play a vital role in adjusting their members' production to the requirements of consumers and improving their economic effectiveness and positioning in the marketplace.
Agricultural cooperatives actively contribute to guaranteeing environmentally-friendly quality products that are made available throughout the whole supply chain.
Agricultural cooperatives are important rural development operators, actively contributing to economic viability in rural areas, including less-favoured regions, by forming and operating the essential information, economic and service-related rural networks, which constitute the backbone of the European social landscape. They are therefore an important source of direct and indirect employment and of economic growth, thus helping to attain the goals of the Lisbon Strategy.
Agricultural cooperatives in the EU are an important socio-economic element in the economy and society at large:
Over 50% share in the supply of agricultural inputs
Over 60% share in the collection, processing and marketing of agricultural products.

Source Copa-Cogeca (2020)

Large number of international projects initiated and implemented by COGA-COHECA, also a number of position papers from COGA-COHECA prepared in 2020 on (1) Gender Equality Strategy, (2) on perspective on long-term vision for rural areas, (3) on Economic Recovery and Job Creation, (4) on Deployment of Seasonal Workers from the European Countries to the EU have demonstrate that COGA-COHECA is an important actor in shaping and further developing the European Union policies that create important framework conditions for cooperative enterprises. COGA-COHECA is an active member in the policymaking process for the EU agricultural policy. Lithuanian Association of Agricultural Cooperatives 'Kooperacijos kelias' is also a member of COPA-COGECA to reflect the needs of the Lithuanian agricultural cooperatives at the international level.

6.5.2 *Case Study 'Lithuanian Association of Agricultural Cooperatives' 'Kooperacijos Kelias'*

1st stage: Emergence of social movement 'Lithuanian Association of Agricultural Cooperatives' 'Kooperacijos kelias'. In 2000, Prof. Antanas Stancevičius and his fellow thinkers founded the Lithuanian Association of Agricultural Cooperatives 'Kooperacijos kelias'. The most important idea which inspired the founding of the Association was to take the advantage of opportunities presented by cooperation and improve the lives of rural people. Agricultural cooperation is one of the most important measures determining the efficiency of farmer's activity. Cooperation provides social and economic benefits, particularly for very small and small farms, helping them to reduce production costs, increase the productivity of agricultural activities and thus to significantly contribute to social and economic vitality of local rural areas.

The main objective of the Lithuanian Association of Agricultural Cooperatives 'Kooperacijos kelias'has always been to expand the cooperation in line with the country's strategic priorities of agricultural development. The Association works to achieve this objective through the implementation of the following major activities: representing cooperative relations with public authorities, domestic and foreign markets partners; seeking credit resources in the markets; introducing advanced production technologies and scientific recommendations, which would significantly increase labour productivity, improve product quality and reduce production costs, thus allowing agricultural products to better compete in the global market; initiating the establishment of new cooperatives, coordinating their activities and preparing strategies for their participation in the markets; organising seminars, conferences, fairs, exhibitions; preparing methodical recommendations and newsletter on the issues relevant to the Association and its members; organizing professional development, training, consulting opportunities and internships in Lithuania and abroad for the members of the Association; collaborating with similar organizations operating in Lithuania, other countries and internationally; providing support and establishing contacts with other entities working in the sector of agriculture, collaborating with the Chamber of Agriculture of the Republic of Lithuania, the Lithuanian Agricultural Advisory Service and other advisory services; participating in the preparation and consideration of new draft laws and other legal normative acts of the Republic of Lithuania, which have impact on the activities of the Association and its members; creating databases on agricultural cooperation and providing these data to the members of the Association and other institutions.

2nd stage: Coalescence of social movement 'Lithuanian Association of Agricultural Cooperatives 'Kooperacijos kelias'. According to the Chairman of the Association 'Kooperacijos kelias', Jonas Kuzminskas, the main link uniting members is the goal to increase the bargaining power with traders, as farmers, especially small ones, experience financial losses while selling their production individually. Traders also understand that it is easier for them to communicate with one seller representing farmers rather than deal with each farmer individually.

Membership in the Lithuanian Association of Agricultural Cooperatives is open. The Association currently unites 12 cooperatives.

The following entities are involved in Association's activities: agricultural cooperative 'AGROLITAS' which unites producers growing vegetables in greenhouses (three legal entities and five individual persons); cooperative enterprise 'Baltic Cattle', which aims to help farmers to sell animals at the highest market prices, as well as to acquire breeding animals of high genetic value; cooperative 'Daržovių centras'; Cooperative 'EKO tikslas', which brings together 48 organic farms and farmers (members of this cooperative own in total about 5,000 hectares of organic arable land area, ecologically farm around 1200 cows with estimated offspring of 450 units, and beef cattle considered to be about 250 units); cooperative 'EKO Žemaitija', which unites ecological dairy farms; agricultural cooperative 'Juodoji Uoga', which unites the largest blackcurrant growers; agricultural cooperative 'Lietuviško ūkio kokybė ', which unites farmers trading in mobile farmers' markets; cooperative 'Maldutis' providing agricultural services; agricultural cooperative enterprise 'Pamario pienas' purchasing milk and providing other services to farmers; agricultural cooperative 'Pienas LT', which buys and processes milk, and two cooperatives 'Pienininkai' and 'Pieno gėlė' which both buy milk.

The members of the Association are actively involved in its activities, because everyone is interested in the concept of cooperation. As a result, the turnover of members is not large. Since 2017, three new members have joined the Association.

3rd stage: Biureaucratization of social movement Lithuanian Association of Agricultural Cooperatives 'Kooperacijos kelias'. The resources of the Association are used to finance the dissemination of information, organize targeted events, and finance international activities and to meet other administrative needs. However, in order to develop cooperation activities, it is becoming necessary to inform the public about the benefits of cooperation, to train and educate the younger generation, and this requires additional financial resources.

Lithuanian Agricultural Cooperative Association 'Kooperacijos kelias' is a non-governmental and non-profit organization, established by legal persons on a voluntary basis. The Association operates under the Lithuanian Republic Law on Associations and other legal acts. The main governing bodies of the Association are the General Assembly of Members, the Board of the Association, the Auditor and the Chairman.

Membership in the Lithuanian Association of Agricultural Cooperatives is open. A member of the Association has the right to participate and vote at the General Assembly of the Association's members, elect and be elected to the Association bodies, use services provided by the Association, get acquainted with Association's documents and receive all information available about its activities. The rights of members are set out in the statutes. Members of the Association must pay the admission and annual membership fee, and cannot perform activities that would harm the interests of the Association and other members of the Association. Members must comply with agreements and obligations between members of the Association, with the statutes of the Association, implement resolutions agreed by the General Assembly, and actively participate in the activities of the Association.

Association's everyday activities are organized by its employees. The Association employs an International Relations Coordinator-Administrator. The Chairman of the Association represents the Association in meetings with the Lithuanian and foreign institutions, and partners. Nevertheless, each member of the Association contributes to the smooth implementation of the activities of the Association.

The Association annually organizes the General Meeting of Members and Board meetings. At least once a year an international conference is organized, and decision makers from Lithuania and the European institutions are invited to participate, also the Lithuanian government members, representatives of foreign cooperatives and representatives of agricultural organizations and farmers in Lithuania take part in the event. Information on past events and official decisions of the Association are uploaded on the website and Facebook account. These activities help to project the image of the Association and strengthen its role as an active player, which disseminates and promotes ideas of cooperation both in Lithuania and abroad.

Daily communication is done by e-mail and telephone. Meetings of the Board of the Association are organized at least 4 times a year to ensure internal communication. The Association is a permanent member of many working groups organized by the Lithuania government, the Seimas, and Ministries. The Association regularly participates in the meetings of the Committee on Rural Affairs and other committees at the Seimas, attends various meetings of the Government and inter-institutional meetings. All the latest background information relevant for members is uploaded at the 'News' section of the Association's Website and Facebook account. The information is constantly updated. Scientists, politicians and government representatives are invited to the meetings of the Presidium of the Association.

Since 2004, 'Kooperacijos kelias' has been a member of COPA (Committee of Professional Agricultural Organizations of the European Union) and COGECA (General Confederation of Agricultural Cooperatives of the European Union). The purpose of this membership is to pursue the Common Agricultural Policy favourable to the Lithuanian agricultural cooperatives, farmers and rural residents, to participate in legislative processes and to actively promote the interests of the Lithuanian agricultural cooperatives at the international level.

4th stage: Decline of social movement 'Lithuanian Association of Agricultural Cooperatives 'Kooperacijos kelias'. Almost five hundred cooperatives have been established in Lithuania mostly in agricultural and food value chains. However, the idea of cooperation is not spreading fast enough. As the European Union's requirements in the agricultural sector increase, cooperation is important in order to remain competitive. It is also becoming increasingly important to educate the younger generation on the benefits of cooperation. At present, we try not only to continue the existing activities, such as representation of the interests of cooperatives both at the national and international level, but also to disseminate examples of good practices and to use social networks to show the benefits of cooperation and attract various stakeholders and young people.

We believe that good examples of cooperation, continuous education and adequate funding can encourage cooperation and, at the same time, help to solve multiple issues in the agricultural sector, and increase the competitiveness of agricultural producers and other stakeholders in the markets.

6.6 New Role of Peasantry

6.6.1 The New Role of Peasantry and Food Sovereignty as Key Innovation Around the World

The last decade of the twentieth century resulted in significant shifts in changing the understanding of the role of peasantry. Oppositely to the prediction of capitalism advocates, who proposed the peasantry to demise and disappear due to the highly struggled agricultural productivism (Martínez-Torres & Rosset, 2010), peasant communities managed to keep running their way of living and alongside improving their quality of life. Nearly 1980s, the governments started less intervening rural areas and agriculture around the world with insignificant timeline differences, and overall situation in rural areas became more and more complicated. Particular active national peasant groups began forming cross-border ties with transnational organizations, starting from Latin America and with the rest part of the world after (Martínez-Torres & Rosset, 2010). The so-called 'new rights advocacy' of the 1990s accelerated the raise of the peasant rights in line with the human political and civil rights' movements, followed by social and economic rights' movements. The spatial newly raising initiatives of people, living in rural areas in different parts of the world, began issuing the common idea of the necessity to understand peasantry as alternative to existing hegemonies, which had been artificially established by profit-minded actors in multistoried business and government structures, and factories. Peasantry started being treated as remaining 'most systematic and comprehensive organic and living alternative' (Patel, 2006). In the late 1990s, peasant communities start responding to the last phase of neoliberal capitalism in evidently more organized and sophisticated way, crossing the national borders (Martínez-Torres & Rosset, 2014).

Among the most important and powerful transnational social movements in the world in the field is an international peasants' movement 'La Via Campesina' (Martínez-Torres & Rosset, 2010). Women and men, farmers' representatives from the four continents, gave birth to the La Via Campesina social movement in 1993, Mons, Belgium (La Via Campesina, 2020). At that time agricultural policies and agribusinesses were increasingly becoming globalized. Aiming to have the smalls' needs to be heard in the decisions which directly affect their life, small-scale farmers should unite, develop their vision and struggle to defend it. La Via Campesina fights for the three key struggles (La Via Campesina, 2020):

1. *Defending food sovereignty, struggle for land and agrarian reforms.* 'Food Sovereignty' political vision was launched by La Via Campesina at the World Food Summit in 1996. Food Sovereignty is understood as the right of people to healthy and culturally appropriate food which is produced sustainably, as well as peoples' right to define their own most appropriate food and agricultural systems.

The vision leads to the development small-scale sustainable production benefitting communities and their environment model. Local food production and consumption is considered as the priority in food sovereignty—it allows the country to protect its local producers from cheap imports and help control production. The struggle for land and genuine agrarian reform help ensure the rights to use and manage lands, territories, water, seeds, livestock and biodiversity that are in the hands of actual producers, not the corporate sector.

2. *Promoting agroecology and defending local seeds.* La Via Campesina believes that the key form of resistance to profit-driven economic system is the agroecology. Half of the world is composed of the small farmers, including peasants, fisher folk, pastoralists and indigenous people, who are capable to produce food to their communities and feed the world in sustainable and healthy way. Irreplaceable pillars of food production are seeds, which at the same time are the basis of productive social and cultural reproduction. La Via Campesina promotes the farmers' rights to use, develop and reproduce peasant's seeds and struggle against attempts by corporations to control the common heritage.
3. *Promoting peasant rights and struggle against criminalization of peasants.* Globally peasants are affected by increased displacement, criminalization and discrimination. Basic rights are violated by transnational corporations, while people are struggling to defend the rights of their communities that continue to become criminalized or even killed. The Universal Declaration on the rights of peasants and other people working in rural areas, promoted by La Via Campesina, includes right to life and adequate standards of living, the right to land and territory, to seeds, information, justice and equality between women and men.

Currently La Via Campesina movement unites about 200 million peasants, small and medium size farmers, landless people, rural woman and youth, indigenous people and agricultural workers from 182 local and national organizations allocate to 81 countries worldwide: Africa, Asia, Europe and the Americas (La Via Campesina, 2020). La Via Campesina social movement defends peasant agriculture to serve food sovereignty by promoting the social justice and dignity and strongly contradicts corporate-driven productivist agriculture, which destroys social relations and nature. The movement is based on a strong sense of unity and solidarity among the groups.

La Via Campesina's as mass movement's vitality and legitimacy come from peasants' organizations at the grassroots. The key organization feature is decentralization of power among regions. Rotation principle under the collective decision, made in International Conference every four years, is applied to the international La Via Campesina secretariat. The International Conference is the broadest for the political room for discussions and decisions regarding the future actions and agenda of the movement. First international secretariat had been located in Belgium (1993–1996), then in Honduras (1997–2004), Indonesia (2005–2013), and is currently based in Harare, Zimbabwe since 2013 (La Via Campesina, 2020).

With such a powerful mass united by La Via Campesina movement, which is broadly scientifically considered as a viable social movement of most innovative rural actors who are able to set prospective ideas for political and social rural development agendas, peasantry is strongly advocated at global level. In political debates and scientific discussions, peasantry gains the new value next to questions of the food sovereignty innovation, which worldwide is more and more often articulated next to sustainability and reliable resilience recipes for future human beings.

The rural or village action movements, as we now know them, began in Finland in the 1970s, in response to rural change and rapid depopulation. The idea has since spread to all of the Nordic countries and most of the EU accession countries. The longest established are in Finland, Denmark and Sweden. Similar organizations developed independently, but for similar reasons, in other parts of Europe, especially during the 1980s. These organizations have evolved and developed to meet the challenges facing rural communities within the differing national contexts. Further surge of activism emerged in Europe in a form of rural parliaments at the beginning of the twenty-first century. Rural parliaments are an expression of a social struggle with no obvious clear-cut protagonist (Šoster & Halhead, 2011). Small farmers, intellectuals, inhabitants of rural areas, workers and entrepreneurs, young people and adults, all of them are represented in rural parliaments. Particular roles and interests are interlinked and the views of specific social groups are often in conflict. In most cases, rural parliaments are events lasting a few days, gathering various stakeholders and respected politicians. Rural parliaments are similar to the organizations from which they sprang. They gather rural people regardless of their religious, political or class affiliation. The demands of rural parliaments include looking beyond the trade union interests of different regional, national and European farmers' organizations. The common aspiration of all actors involved in rural parliaments is to protect the interests of rural areas. This leads towards a territorial rather than class-based positioning of the social movements involved, a general feature of globalization. Rural networks, initiators of rural parliaments, represent an important part of the new social movements, moving from urban areas to rural ones, calling to revise peasantry/society relationships according to food, energy and technological sovereignty concept and suggest new ways of the optimal organization of society by transforming all spheres of life.

Rural parliaments are regular annual or biannual gatherings in five European countries: Sweden, Finland, the Netherlands, Estonia and Slovakia (Šoster & Halhead, 2011). Four of them are organized by Prepare member organizations. In addition, other Prepare member organizations have expressed an interest to apply the methods or events called 'rural parliaments' in their countries. The European Rural Parliament was established in is a long-term campaign established to express the voice of rural people in Europe, and to promote self-help and action by the rural people, in partnership with civil society and governments (Šoster & Halhead, 2011). The European Rural Parliament is co-initiated by three pan-European networks:

- European Rural Community Alliance (ERCA);
- PREPARE Partnership for Rural Europe;
- European LEADER Association for Rural Development (ELARD).

A major role in the European Rural Parliament campaign is being played by European and national partners, most of whom are members of the co-initiating networks, including Lithuania.

6.6.2 Case Study: Social Movement 'Lithuanian Rural Parliament'

1st stage: Emergence of social movement 'Lithuanian Rural Parliament'. The idea to establish Lithuanian Rural Parliament (hereinafter referred to as the LRP) arrived from abroad practices in rural areas, especially from Finland and Estonia, and was accelerated by active people in

Lithuanian Union of Rural Communities. The LRP should bring together active rural people from all over Lithuania with one basic goal: to discuss the emerging opportunities of Lithuanian rural development and to find new solutions.

The initiative was realized in 2014. The Lithuanian Union of Rural Communities (hereinafter referred to as the LURC), together with the Local Action Group Network (LAG Network), in partnership with Vytautas Magnus University Agriculture Academy (that times it was Aleksandras Stulginskis University—the main Lithuanian university for agriculture and rural development studies in Lithuania), convened the 1st founding session of the LRP on 11 April 2014 (Lithuanian Rural Parliament, 2020) in Kaunas, the geographical centre of Lithuania.

Taking into account the general tendencies of rural development throughout past years, the participants of the founding LRP session in 2014, stated that the central government focuses exclusively on commercial agricultural and food production and solving the problems in this sector, and the main issues of Lithuanian rural demography, rural development remain forgotten. In rural areas, the population is threateningly declining, new businesses are not being created, jobs are shrinking, rural schools, outpatient clinics, libraries and other public service institutions are being closed, and a huge lack of pre-school education and social services is observed. Young people are not encouraged to work and build their future in the Lithuanian countryside.

Thus the initial LRP in 2014 appealed to the institutions of local self-government, the Lithuanian state and the European Union and called for immediate attention to the country's rural problems and for real, rather than declarative, measures to be taken with the help of active Lithuanian rural people.

In its initial session April 11, 2014, the 1st session of LRP approved a Resolution on 'Improving the quality of life in Lithuanian rural areas' (2014), which stated the inconsistent living standards in Lithuanian rural areas, and addressed key issues, threats and concerns of rural community for passive role of responsible actors in the field.

The power, accumulated by the 1st LRP session, was directed to accelerate discussions addressing seemingly forgotten Lithuanian rural people and their problems. Thus the 1st LRP Parliament addressed the 1st Resolution to the key governance institutions, responsible for rural policy formation and implementation at the national level and broader:

- The President of the Republic of Lithuania;
- Committee on Rural Affairs of the Seimas of the Republic of Lithuania;
- To the Government of the Republic of Lithuania;
- Ministry of Social Security and Labor of the Republic of Lithuania;
- Ministry of Agriculture of the Republic of Lithuania;
- Ministry of Education and Science of the Republic of Lithuania;
- The European Commission;
- The European Parliament

The initial propositions evidently signalized the existing mismatch between the expectations for living quality in Lithuania, compared to the whole country, regional centres and rural areas. The 1st LRP approved a few outstanding concerns and proposed suggestions for solving the existing problems, as outlined below.

Promoting community spirit and volunteering. Communities are the most active form of civic initiative. Rural community organizations in Lithuania operate in accordance with the Law on Associations, which greatly complicates activities. The community is not just a simple NGO. Fostering volunteering and strengthening the authority of the volunteer must start at the school bench. Each member of the community must first be involved in volunteering and this must be reflected in the volunteering contract that must be concluded with each volunteer. Communities, LURC and LAGs should organize annual volunteer elections. The family must cherish the values of volunteering, community and good neighbourhood. If you do not participate in the activities of at least one public organization, you are a poor citizen. Promoting and supporting citizenship is one of the most important tasks of a community organization. Citizenship education campaigns, commemoration of state and national holidays, motivation of active community citizens must find a place in the activities of each community organization. In order to strengthen community spirit and volunteering, there had been given a few propositions to the Government of the Republic of Lithuania: first, provide state tax benefits to ensure the status of volunteers (e.g. tax exemptions for at least 500 hours per year as a volunteer, etc.); second, simplify accounting for voluntary organizations; third, improve the legal framework in line with community principles.

Implementation of the LEADER method. The natural mission of the community is to take care of the people of its village, town, city, to mobilize them and to initiate the development of social and physical infrastructure. In recent years, the use of EU support under the LEADER method has shown the results of successful rural communities: renovated community houses, recreation areas, landscaped parks, stadiums, restored churches, community cooperatives, craft centres, etc. It can be confidently stated that communities in many cases significantly increased the assets of the state and local government. Many communities participated in the implementation of the Rural development programme 2007–2013 measures under the LEADER method and other international and national programmes. The LEADER approach has been a strong motivator for community-based bottom-up action, but there is concern about the growing top-down trend in some rural areas and at national level. Irreparable damage has already been done by the fact that Lithuania rejected in the 2014–2020 programming period a community-based local development model, which was proposed by the EU Commission for the period 2007–2013, and which was to combine urban and rural community initiatives and strengthen urban–rural integration links. Following the situation, LRP suggested 8 proposals to be made by the Government of the Republic of Lithuania to maximize the benefits of the LEADER approach.

Strengthening rural territorial self-government. The main self-governing organization in the village is the community organization. Elders are elected in those rural settlements where no community organization has been established. When a community organization is formed in a rural area, the elder's representation is destroyed. In granting the status of representation of local residents to the rural community, LRP proposes in the Law on Local Self-Government of the Republic of Lithuania to legalize the duties and responsibilities of the community chairman and the composition, duties and responsibilities of the community collegial management body. Local municipalities should find ways to cover the minimum costs of community activities from municipal budgets.

Problems of rural youth. The vitality and survival of the countryside depend directly on the people living, working, young people starting families. The number of young people in Lithuanian villages and towns is drastically decreasing. Every year, Lithuanian rural areas lose up to 5% on average young people of working age. In today's Lithuanian village, young people see no prospects of staying and creating family, raising children in rural areas where there are neither schools nor kindergarten, neither living space nor work. LRP proposed 3 ways of solving these problems: first, to create opportunities for young families in villages and towns to get living space or to grant a preferential right to acquire a plot for the construction of a residential house; such support would attract young families to stay and create in the countryside; second, to create opportunities for young people on preferential terms (for the first two years without making a profit tax and income per capita tax) to set up micro-businesses, etc.; and third, review the education system in rural areas by returning rural schools to 7-year-olds or primary school statuses; by closing schools in villages and towns, since rational use, the state's financial resources do not pay off in the long run: when rural children leave the countryside in early age and go to urban schools, they tend no longer to associate their future with rural areas.

Business creation and development in rural areas. The total lack of support for starting self-employment initiatives or other small entrepreneurship in rural areas had been observed. 11 measures had been proposed to accelerate changes in the business creation situation in rural areas, including tax reduction for new businesses; tax benefits for newly established workplaces or self-employment; one-time support (e.g. 10 thous.) for starting the micro-business activity in rural areas; EU support allocation prioritized by the creation of small cooperatives, social enterprises providing services to micro-enterprises in rural areas; preferential treatment for small shops or outlets, trading in local products, located in administrative centres, or larger in the towns of the district; and many other preferences for starting the business in rural areas.

Social services in the countryside. Social service institutions are very unevenly distributed in the districts of Lithuania. Almost all social services are concentrated in cities, although most people who need it live in rural areas. It is necessary to encourage the initiative of communities to develop a network of social services in rural areas. It is becoming increasingly difficult for rural people to access basic services: primary schools, shops are closed, childcare facilities, post offices, railway stations and other miscellaneous facilities. Under such conditions, the threat of social exclusion increases. Although many Lithuanian Government documents note decentralization of social services, in practice, there are limitations to the various legal instruments for providing social services. There is no appropriate mechanism for NGOs and local authorities to work together in organizing social services in rural areas. The demand for social services is huge in rural areas: lonely elderly people, the disabled, families with social problems, deprived of parental care children, orphans, returnees, alcohol or drug abusers and other social groups, so the availability and quality of social services play vitally important role. Adaptation of particular infrastructure and maintenance of services is a continuous challenge for rural community organizations, at the same time—new impetus for rural development by creating new jobs and reducing social exclusion in rural areas. Accordingly, appropriate measures had been proposed by LRP to solve the existing situation to the Government of the Republic of Lithuania, to the territorial self-governance and to the communities.

The emergence of LRP social movement was crowned with strong and much promising statement, which demonstrated the high ambition and guiding cornerstones of the movement: 'Only strong community organizations can ensure the active participation of communities in the country's social, economic and political life and the development of civil society' (1st Resolution of Lithuanian Rural Parliament, 2014).

2nd stage: Coalescence of social movement 'Lithuanian Rural Parliament'. The emergence of LRP movement in Lithuania was followed by the coalescence, accelerated and advocated by indifferent people from Lithuanian rural society, including highly educated academics (Vytautas Magnus University Agriculture Academy), civil society organizations (LURC and LAG Network) and many active community members, who highlighted the initially stated concerns of rural future in Lithuania and were able to propose multiple solutions to move forward.

Since the establishment in 2014, LRP has been organized 4 times already. All rural community leaders, politicians, business makers and scientists are invited to come and take part in parliament sessions. Around 400 participants took part in the first and the second parliaments, the latter two Parliament sessions had been attended by nearly 150 people. There is a natural change in membership of the Parliament, since the new active people come to rural areas with innovative ideas, striving for new solutions which are necessary for rural areas to develop and prosper current times in knowledge age.

The main idea of the parliament is to unite as many as possible different people from different professions and background, caring about issues of rural development in Lithuania, to issue and summarize the main existing concerns of rural areas in Lithuania naturally.

3rd stage: Bureaucratization of social movement 'Lithuanian Rural Parliament'. Structure of the movement and main roles. The main LRP organizational activities are done by LURC active members in partnership with Vytautas Magnus University Agriculture Academy, which normally provide the venue for the Parliament session (except the 3rd session, which was organized in another Ukmergė district, Taujėnai manor) and the LAG Network. LRP active members are leaders from 40 LURC umbrella organizations in various regions of Lithuania with 1229 unique rural community organizations.

The roles of LRP members are distributed as follows. There is a work group of six people (consisting of two people and a board member of each organization) and the chair of the LURC, which organize LRP sessions. The work group manages the entire event—LRP session: they decide on the agenda and the people who are invited to come to talk about their experiences. The chair is in touch with the politicians, sponsors, scientists and managing authorities, and is also in charge of the practical aspects of event organization.

The topics to be discussed during the LRP sessions are formulated before the sessions. Topics are defined during the decentralized meeting around the country, organized by LURC representatives in different regions together with active rural leaders of that region. After such meetings, the topics are summarized and the project of Resolution is prepared and spread among the participants of coming LRP. Then topics are presented by invited responsible speakers and finally ERP votes for approving the Resolution which is to be provided to the list of recipients according to the fields of topics discussed. Approved Resolution is further forwarded to recipients as well as the international organization—European Rural Parliament.

Resources. Overall organizing activity of LRP is based on voluntary work basis. Qualified people had already matured among the members of LURC and the LRP, therefore organizing and other administrative work is done using the internal human resources. Financial resources are attracted from outside in a form of private sponsorship or project funding from different national programmes and other calls for applications. The first LRP was funded by a private sponsor with a budget of 900 EUR and proved to be a success. Further LRP sessions were financed under the application made to the national funds, such as the Lithuanian Rural Network (e.g. support for the 2nd LRP session amounted 3,000 EUR). The experience gained by organizing the parliament sessions was very important and useful both for organizers of LRP, as well as for members of LURC.

Identity of the movement. The LRP is well known among the people in Lithuania who are not indifferent regarding the rural development issues in Lithuania. LRP is recognized as part of LURC activity rather than an independent movement. The movement has an identity of opposing and proposing force, which cares about the future of Lithuanian rural development and is able to formulate propose consolidated solutions at regional, national level, as well as cross-national boarders. LRP is also well known at the Lithuanian government level as all LRP sessions include invited representatives from policymakers and the Government of the Republic of Lithuania, aiming to make the voice of rural people from rural communities heard and reflected.

Communication channels. LRP uses internal and external, formal and informal communication channels. Internally, each LRP session is announced via LURC regional representatives in advance, aiming to invite active rural people to discuss the main issues of rural areas, which should be outlined in the Resolution of the Parliament session. The invitation to attend the LRP session is open and is normally announced via the LURC website. Invited speakers are connected directly by organizers. Every LRP session is reflected in mass media, local and regional newspapers or newsletters, normally in a form of short interview with organizers and the chairman of the LURC (currently—Virginija Šetkienė). All four LRP Resolutions are free accessible in an open source—LURC website's separate worksheet (Lithuanian Rural Parliament, 2020), devoted to LRP.

International activities. The role of LRP greatly increased due to the participation in European Rural Parliament. The resolutions, approved by LRP, are translated into English and provided to the European Rural Parliament, which is added to the particular document—Manifesto of European Rural Parliament and policy suggestions regarding the most recent issues of the development of Rural areas in Europe. Thus the voice of LRP is reflected in the overall European Rural Parliament opinion, since the campaign reaches a climax every two years with a gathering of rural people from all parts of the wider Europe.

4th stage: Decline of social movement 'Lithuanian Rural Parliament'. The decline of the Lithuanian Rural Parliament might be characterized as co-optation for the common activity of rural development indifferent people. Despite the fact, that the first two sessions of the Lithuanian Rural Parliament united more participants, the latter two sessions already demonstrate particular matureness of the movement. The LRP had already developed internal leaders from the movement who are able to organize and moderate sessions, whereas priorly it was normally done by external services. At the same time, it should be stated, that LRP had not become a mainstream yet with well-heard and perceiving placed suggestions to solve the existing rural issues. It becomes evident from the latter Resolution (2019) declaration, which states the continuity of problems, raised in the first LRP:

1. that while implementing 2014–2020 LEADER/CLLD program, which had the goal to decrease poverty and social separation in villages, in local development strategies the spirit of LEADER method disappeared, local community gathering and empowerment projects did not receive enough funds and that is why a big part of community organization activities started decreasing and the set goals are not implemented. The main funds (around 75%) are directed toward business and creating workplaces and even though community and/or social business creation is supported, a lot of community organizations are not able to take up these obligations because of lack of human, financial and other resources;
2. that too long periods for program development and approval, publication of project tenders and evaluation of project applications and payment claims, especially evident when implementing LEADER/CLLD program. Because of prolonged procedures, the community organizations become an unreliable client/buyer, who cannot pay the suppliers, implement the planned project activities/tasks on time;
3. that the surplus support fund usage regulation negatively affects village leaders' activity, because community organization leaders are worried about "from top received" and hard to implement project obligations, surplus document demand when applying, difficult decision-making process in regards to giving funds for the projects;
4. that only cross-ministerial integrated measures to strengthen the human and other resources of local communities can address the current situation. The development and implementation of multi-fund local development strategies would greatly help to this end.

Participants of Lithuanian rural parliament are addressing Lithuanian state institutions and are urging them to give more attention to solving these urgent problems. This gives evidence, that the voice of LRP had not been responded to in rural policy according to their wills and ambitions. Indeed, the LRP, as stated by the LURC chair Virginija Šetkienė, cherishes a great hope to get heard one day and to arrange favourable and flexible conditions for modern rural development in Lithuanian regions in collaboration with the policy makers and the overall rural development support system of the EU.

References

Agricultural Information and Rural Business Center. (2020). https://www.vic.lt/valdos-ukiai/statistika/ukininku-ukiu-statistika/. 16 November 2020.

AIC, Agricultural Industries Confederation. https://www.agindustries.org.uk/. 19 November 2020.

Andrews, M. (2020, October 30). *The rise of rural women's movements in Southern Africa*. https://www.cultivatecollective.org/in-perspective/the-rise-of-rural-womens-movements-in-southern-africa/. 16 November 2020.

CELCAA, the European Liaison Committee for Agricultural and Agri-Food Trade. https://celcaa.eu/.

Copa-Cogeca (2). (2020). https://en.wikipedia.org/wiki/COPA-COGECA. 26 November 2020.

COPA-COGECA History. Official website of the organisation. https://copa-cogeca.eu/CogecaHistory.aspx. [23 November 2020.

COPA-COGECA, Committee of Professional Agricultural Organisations - General Committee for Agricultural Cooperation in the European Union. (2020). https://copa-cogeca.eu/.

Edelman, M. (2005). Bringing the moral economy back in… to the study of 21st-century transnational peasant movements. *American Anthropologist, 107*(3), 331–345.

ELO, European Landowners' Organization. https://www.europeanlandowners.org/. 16 November 2020.

Euroseeds. https://www.euroseeds.eu/. 26 November 2020.

Hoggart, K., Black, R., & Buller, H. (2014). *Rural Europe*. Routledge. https://www.globalagriculture.org/fileadmin/files/weltagrarbericht/Weltagrarbericht/10B%C3%A4uerlicheIndustrielleLW/Pocketbook2018.pdf. 16 November 2020.

Index of Economic Freedom. (2020). https://www.heritage.org/index/property-rights. 16 November 2020.

Jose, A. V. (2013). Changes in wages and earnings of rural labourers. *Economic and Political Weekly, 48*, 107–114.
Klupšas, F. *Lietuvos žemės ūkis*. https://www.vle.lt/straipsnis/lietuvos-zemes-ukis/. 26 November 2020.
La Via Campesina. (2020). International Peasants' Movement. https://viacampesina.org/en/. 14 September 2020.
Lithuanian Family Farmers' Union. http://seimosukiai.lt/. 14 August 2020.
Lithuanian Rural Parliament. (2020). http://www.lkbs.lt/index.php?277450449. 19 August 2020.
1st Lithuanian Rural Parliament Resolution "Improving the quality of life in Lithuanian rural areas". (2014, April 11). http://www.lkbs.lt/index.php?277450449. 20 August 2020.
4th Lithuanian Rural Parliament Resolution. (2019, October 18). http://www.lkbs.lt/index.php?277450449. 20 August 2020.
Mathijs, E., & Swinnen, J. F. (1998). The economics of agricultural decollectivization in East Central Europe and the former Soviet Union. *Economic Development and Cultural Change, 47*(1), 1–26.
Martínez-Torres, M. E., & Rosset, P. M. (2010). La Vía Campesina: The birth and evolution of a transnational social movement. *The Journal of Peasant Studies, 37*(1), 149–175.
Martínez-Torres, M. E., & Rosset, P. M. (2014). Diálogo de saberes in La Vía Campesina: Food sovereignty and agroecology.". *Journal of Peasant Studies, 41*(6), 979–997.
Patel, R. (2006). International agrarian restructuring and the practical ethics of peasant movement solidarity. *Journal of Asian and African Studies, 41*(1–2), 71–93.
Petras, J., & Veltmeyer, H. (2001). Are Latin American peasant movements still a force for change? Some new paradigms revisited. *The Journal of Peasant Studies, 28*(2), 83–118.
Platform Lithuanian Family Farmers Union. https://www.facebook.com/seimosukininkusajunga/ and https://www.kaimasveza.lt/. 15 August 2020.
Robles, W. (2001). The landless rural workers movement (MST) in Brazil. *The Journal of Peasant Studies, 28*(2), 146–161.
Robles, W., & Veltmeyer, H. (2015). *The politics of agrarian reform in Brazil: The landless rural workers movement*. Springer.
Rucht, D. (1996). The impact of national contexts on social movement structures: A cross-movement and cross-national comparison. *Comparative Perspectives on Social Movements: Political Opportunities, Mobilizing Structures, and Cultural Framings*, 185–204.
Šoster G., & Halhead, V. (2011). *Rural parliaments: Emerging participative democracy*. Brussels, Ljutomer: PREPARE Network. http://preparenetwork.org/files/PREPARERuralParliaments.pdf.

Steup, R., Santhanam, A., Logan, M., Dombrowski, L., & Su, N. M. (2018). Growing tiny publics: small farmers' social movement strategies. *Proceedings of the ACM on Human-Computer Interaction, 2*(CSCW), 1–24.

The Ministry of Agriculture of the Republic of Lithuania. (2018). https://zum.lrv.lt/lt/naujienos/zemes-grazinimas-sparciai-arteja-prie-pabaigos. 17 September 2020.

The National Land Service under The Ministry of Agriculture of the Republic of Lithuania. (2020, June 11). http://www.nzt.lt/go.php/lit/Nzt-baige-zemes-paemima-projektui-rail-baltica-igyvendinti-. 17 September 2020.

UIPI, International Union of Property Owners. https://www.uipi.com/. 16 November 2020.

Vidickiene, D., Melnikiene, R., & Gedminaite-Raudone, Z. (2016). The influence of the post-industrial economy on economic cooperation needs and forms. In F. Taisch, A. Jungmeister, & H. Gernet (Eds.), *Genossenschaftliche Identität und Wachstum = Cooperative Identity and Growth* (pp. 495–502). 18 th Conference Proceeding of ICCS 2016 in Lucerne/Switzerland. Raiffeisen Schweiz, St. Gallen. ISBN 978-3-033-05782-1.

WFO, World Farmers' Organization. https://www.wfo-oma.org/. 17 September 2020.

CHAPTER 7

Industrial Rural Development Paradigm Shift-Focused Social Movements

Živilė Gedminaitė-Raudonė and Vitalija Simonaitytė

7.1 New Forms of Agriculture for Changing Farmer/Nature Relationships

7.1.1 *Movement of Agroecology Around the World*

Various types of agroecology practices such as organic farming, biodynamic agriculture, agroforestry, permaculture movements and others are significantly increasing in recent decades as more people decide to try these practices in their daily farming. Agroecology is becoming a global movement. Global agroecology movements have called for 'scaling out.' Scaling *out* agroecology would support farmer-to-farmer exchange that spreads agroecological practices through existing, and expanding, networks of small-scale family producers (Roman-Alcala, 2018). The agroecology movement helps to reduce usage of external inputs and advocates for small-scale family farming as relevant to individuality.

Scale up of the Agroecology movement is an important tool that could ensure implementation of goals of the 2030 Agenda. The 2030 Agenda for Sustainable Development calls for a transformation in food and agricultural systems. The 2030 Agenda including its 17 Sustainable Development Goals (SDGs) and 169 targets was adopted on 25 September 2015 by Heads of State and Government at a special UN summit. The adoption of the 2030 Agenda was a landmark achievement,

R. Vilkė et al., *Rural Economic Developments and Social Movements*, Palgrave Advances in Bioeconomy: Economics and Policies, https://doi.org/10.1007/978-3-030-71983-8_7

providing for a shared global vision towards sustainable development for all (The 2030 Agenda, 2015).

Another important role for this movement is devoted to the Food and Agriculture Organization of the United Nations (FAO). FAO advocates for agroecology practices. FAO highlights (2018) that agroecology brings together scientific disciplines, social concepts and practices. The scientific core of agroecology is based on applying ecological concepts and principles to optimize interactions between plants, trees, animals, humans and the environment. In addition, agroecology places social issues at the centre of solutions for a sustainable and fair food system. As a set of practices, agroecology provides multiple benefits to society and the environment, by restoring ecosystem services and biodiversity. Thanks to its integrated approach, agroecology is a key to boost food and nutrition security, while improving the resilience of agroecosystems (FAO, 2018a).

The important contribution of agroecology towards sustainable development is focused on:

- enhancing smallholder and family farmers' adaptation and resilience to the impacts of climate change;
- improving nutrition including through more diversified diets;
- protecting and enhancing agro-biodiversity in support of ecosystem services such as pollination and soil health;
- improving livelihoods in rural areas;
- achieving a transformative change in agricultural practices towards sustainable development (FAO, 2018b) (Table 7.1).

Movement of Agroecology is spread all over the world and includes various types of practices as organic farming, biodynamic agriculture, agroforestry, permaculture movements and others. The history of agroecology depends on whether you are referring to it as a body of thought or a method of practice, as many indigenous cultures around the world historically used and currently use practices we would now consider utilizing knowledge of agroecology (Johnson, 2012).

7.1.2 *Case Study 'Natural Agriculture'*

1st stage: Emergence of social movement 'Natural agriculture'. The main initiators of natural agriculture in Lithuania were Saulius Jasionis and Laimis Žmuida. Inspiration for such initiative for Laimis Žmuida was values proposed by his parents with focus on nature-friendly practices in

Table 7.1 Key facts of agroecology by FAO

Key facts of agroecology by FAO
Agroecology relies on ecosystem services to improve productivity
The knowledge and practices of farmers and food producers from all over the world are at the core of agroecology
Agroecology can address the root causes of hunger, poverty and inequality
Agroecology combines farmers' knowledge with modern science in innovative ways
Agroecology provides local solutions for global challenges
Agroecology relies on the interactions between plants, animals, humans and the environment, to build sustainable and fair food systems
Agroecological approaches are vital for the challenges we face today and tomorrow
Agroecology uses fewer external resources, reducing costs and negative environmental impacts
Sharing and creating knowledge among food producers is at the heart of agroecology

Source www.fao.org

their daily life. When Laimis had finished his studies, then he has started to search for land in rural areas of Lithuania where he could start farming based on his proposed values. He was reading various books on agriculture principles and proposed it also to his friend Saulius Jasionis who also looked for natural farming techniques that helps to reduce usage of external inputs and advocates for small-scale farming as relevant to individuality.

The first actions to create natural agriculture movement have started in 2008. In the beginning, enthusiasts of natural agriculture in Lithuania relied on N. I. Kurdiumov books, which propagate gardening with no soil work (see reference Kurdiumov, 2012a). Kurdiumov in his books described agricultural systems proposed by other authors. The biggest inspiration for creation of natural agriculture for both initiators was from A. Kuznetsov and F. Gelcer ideas (see Gelcer, 2012; Kurdiumov, 2012b). Though eventually, by experimenting, gathering knowledge and adapting both authors advice to practice, the new unique theory was created of natural agriculture, and suggested the techniques most suitable for Lithuanian climate. Natural agriculture is different from conventional and organic farming because it is based on purely natural processes. In conventional and organic farming soil is continuously being emaciated and destroyed, whereas in nature soil its fertile layer is constantly growing, due to yearly weed harvest. Weeds fall down to the soil, are being decomposed by various microorganism and becomes natural fertilizer and food for the agricultural plants. Surplus is being conserved as humus. Contrary to conventional farming in natural agriculture there is no digging, scratching or hoeing. This lets the soil create its own natural structure, which ensures good air circulation, humidity and wealthy environment for live microorganisms. Processes of nature could be imitated by covering the soil surface with the mulch—old organic materials like dry grass, leaves, sawdust, etc. Bacteria and fungi, which are decomposing organic cover, release carbon dioxide into the soil. CO_2 is heavier than the air, meaning that it is observed by the soil, where it melts down and becomes fertilizer for the plants. By using this method of agriculture, the quality of products as well as the environment is better and healthier. Moreover, eventually, the harvest in such areas should be bigger than in ploughed ones. Finally, there is no need to buy fertilizers, insecticides or pesticides, as the balance of microorganisms in the soil creates barrier for plant diseases and pathogenic organisms.

The main goal of the social movement of natural agriculture was to create a new form of agriculture changing farmers and nature relationships applying nature-friendly principles. When both initiators have tried such initiatives in practice then development of this practice at larger scale have started. More interested people were involved who also wanted to apply natural farming principles in their gardens. The next step was the creation of the website www.gerazemdirbyste.lt as the platform of the movement and the forum of natural agriculture enthusiasts. However, the number of the movement members was rapidly growing, due to the society's interest to use natural and healthy vegetables and fruits. For one part of the members, natural agriculture is beneficial in financial means as to grow your own food is cheaper than to buy. For others, quality of the food was the issue, as it differs if you buy products in shopping centres, markets or grow it by yourself using natural agriculture method. Most of these people not only care for their health and eating habits but also want to live in sustainable manner, to apply nature-friendly principles.

Later, initiators of this movement have started to organize various events and seminars to introduce to this farming method as more enthusiasts were approaching them with many questions and will to start this natural agriculture practice in their farming.

2nd stage: Coalescence of social movement 'Natural agriculture'. During informal virtual communication via platform www.gerazemdirbyste.lt between people who strive to use techniques of natural agriculture, the need for seminars appeared, where interested ones or practitioners could exchange knowledge and share their experiences. Later initiators have started to organize lectures and practical workshops in selected place (for example, village, farming place, meeting place, etc.), with the aim to explain principles of natural agriculture and help to adapt them to the local particularities. One of the initiators of this movement Laimis Žmuida has planned that such seminars and consultations will become his main activity and livelihood.

Collaboration between movement members helped to gather best practises and experiences of natural agriculture as well as to improve the technology, convert theoretical knowledge into practices and to adapt it to Lithuanian conditions. One of the initiators of this movement and the lecturer Saulius Jasionis highlighted that 'people who want to use new agricultural methods are scattered all over the Lithuania, but when they come together they have a possibility to share their experiences. When all the experiences and the best practice examples are put together, analysed and systemized, the result could be achieved much faster. Also, for some people it is more convenient to follow known and examined techniques, than to experiment by their selves. The network helps for them by lectures and practical workshops' (Vidickiene, 2013).

The platform of the movement serves not only for collaboration and communication among users of natural agriculture methods and products, but also as an informational tool for those, who want to buy products of natural agriculture. From 2011, *the commerce system of natural agriculture products was established* and some of the growers have started to sell their products.

Movement members of natural agriculture also have *prepared requirements for the products' certification*, including obligatory growing conditions, strict agro-technique control and products' expiry date. According to natural agriculture certification rules, products should be sold in a period of 24 h, counting from the moment, it was picked. Most of the products are sold in the growing places, when berries, vegetables and fruits are picked only when the actual customer comes. It helps to ensure that the product is the most fresh and of the best quality.

Members of this movement are continuously searching for new attractive ways how to spread the knowledge about natural agriculture. One of the examples of their activities, *'Tasting fest' of natural agriculture products* was organized at the beginning of natural agriculture practices. The idea was suggested by one of the movement initiators Laimis Žmuida. He wanted to select the most delicious potatoes for the future growing and started to taste 13 different kinds of potatoes, grown by him. So he cooked them, fried them, compared the tastes and made notes. 'I loved the process of tasting, so I thought that it would be a good idea to share such experiences with others and to organize a public event'—says Laimis. The first Tasting fest in the restaurant 'Gurmė' situated in Kaunas was organized with doubts, excitement and a bit of fair—what if nobody likes it? This event was successful and participants were excited to taste different kind of potatoes grown by principles of natural agriculture. Because of this reason, these people could easily distinguish the differences between potatoes. 9 different kinds of potatoes were prepared for tasting. Participants were served 9 pieces of different cooked and 9 pieces of fried potatoes. Potatoes were cooked and fried without any salt or oil. It was done on purpose to reveal the real taste of the potatoes (Vidickiene, 2013).

The event of exchanging seeds and sprouts became an annual tradition. The idea of the event is to exchange between network members seeds and sprouts of good quality, examined and grown in their own garden. Sometimes people bring for exchange imported seeds unknown to most Lithuanians. Seeds are the main currency in the market and only in rare cases, you can buy desirable seeds for the real money, as all participants want to exchange their seeds for some new plants for their garden.

Membership of the movement of natural agriculture is informal. Annually one gathering of members is organized to share their experiences with the number of participants approximately around 200. Every year 20–30 lectures and practical workshops are organized with 10–15 participants in each event so potential members of this movement count from 200 to 300 annually. Members are people from cities and rural areas that would like to apply natural agriculture practices in their farming activities. Many interested new members join this initiative with already some knowledge of natural agriculture. Another part of people was inspired by neighbours or friends who already are part of this movement. Most of the members are small farmers, inhabitants of other professions from cities or rural areas who want to have their garden where principles of natural agriculture would be applied. Both females and males, equally, support these ideas and are practicing this type of agriculture. More than 50% of participants are young people, with age less than 35 years. Every year is continuous growth of new members who firstly try to get more knowledge on natural agriculture in practical seminars and then to try this activity in their daily lives. The remaining part of members is stable with informal communication via platform or in annual meetings.

3rd stage: Biureaucratization of social movement 'Natural agriculture'. *Resources of the movement.* Human resources are the most important resources in the movement of natural agriculture. The role of initiators of Saulius Jasionis and Laimis Žmuida are of great importance to maintain this movement in the past, today and in the future. Using human resources, the new unique theory was created of natural agriculture, which supplemented N. I. Kurdiumov's teaching and suggested the techniques mostly suitable for the Lithuanian climate by experimenting, gathering knowledge and adapting his advices to practice.

Other resources:

1. IT service: (a) creation and maintaining of the website www.gerazemdirbyste.lt as the platform of the movement and the forum of natural agriculture enthusiasts; (b) Facebook platform as a tool for information and communication with members of movement;
2. financial resources to organize seminars, workshops and events. The amount is not high as the initiator of the movement Laimis Žmuida is searching for supporters of natural agriculture who could provide their settlement as a place for seminars or workshops and usually, expenses are very low. No additional resources are needed at this stage of development.

Structure of the movement. The structure of the movement is decentralized informal network. Website and Facebook platform are used for communication among members. Annually one gathering of members is organized to share experiences of members with the number of participants approximately around 200. Membership in the movement is open to any person who supports ideas of natural agriculture and wants to apply this practice in their garden.

Roles of the movement. The main role from the establishment of the movement was provided by both initiators. Laimis Žmuida is responsible for organizing events and practical workshops and spreading this information on the website www.gerazemdirbyste.lt and Facebook platform. Roles for other members depend on event, practices that are going to be organized and are discussed beforehand.

Identification of the movement. Natural agriculture is becoming more popular in Lithuania every year since establishment of the movement in 2008. Members of the network are the main actors that spread information about natural agriculture in Lithuania using various channels as platforms, TV broadcast, websites that help to create the identity of the network. The movement covers the whole area of Lithuania as an enthusiast of natural agriculture are from different places from Lithuania in rural areas and places close to cities (as community gardens, city gardens, etc.). The traditional event is organized every year as the annual meeting involving about 200 participants.

Communication channels. Platforms www.gerazemdirbyste.lt and Facebook are tools used for (1) internal communication among members of the network; (2) spread information about the movement to the wider society; (3) tool to organize work. Both platforms are used very actively. Initiators of movement are responsible for maintaining the work of both platforms.

Members of the network also participate in some other movements as ecovillage movement where members of the movement also support ideas of nature-friendly practices, ecological way of living with no impact on environment, etc.

4th stage: Decline of social movement 'Natural agriculture'. The current stage of development of social movement 'Natural agriculture' is successful with the prospect for bigger growth in the future. From the beginning in 2008 until 2020 every year new enthusiast participates in lectures and practical workshops on natural agriculture with 400 to 600 participants every year. Initiators of the movement have no intention for bigger growth and they let movement develop in natural way. If there would be a need for more participants in this movement, additional resources would be needed as (1) education of society on nature-friendly ways to live with nature, new form of agriculture changing farmer and nature relations, etc.; (2) new programmes in universities and at schools; (3) new requirements for farmers using the EU and/or national support, etc.

This social movement has an impact to the society by proposing (1) new form of agriculture changing farmer and nature relations; (2) advocating of new small-scale family farming as relevant to individuality, (3) creating and/or strengthening new food consumption habits and requesting new quality of food.

7.2 New Quality of Food for Changing Customer/Food Relationships

7.2.1 *Food and Consumers' Movements Around the World*

Food movements are rapidly growing social and political phenomena almost all over the world. In many countries, there have been surges of interest in heirloom seeds, in craft beers, in traditional bread and baking, in the demand for city garden plots, in organic food and in opposition to GMOs. Simultaneously, there has been a massive growth of interest in food on social media and the initiation or renewal of initiatives such as Slow Food and many others (Latham, 2016). There are a huge amount and even bigger variety of food consumers' movements across the world, an absolute majority of them are fighting for better, more sustainable, cleaner, and fairer food. Food consumers' movements can be divided into several groups: slow living, slow food and down-shifting; local, organic, sustainable and fair food; and finally food that does not contain dangerous chemicals, pesticides and genetically modified organisms (GMOs).

Slow Food International is one of the worldwide known global, grass-roots organization, founded in 1989 to **prevent the disappearance of local food cultures and traditions, counteract the rise of fast life and combat people's dwindling interest in the food they eat,** where it comes from and how people's food choices affect the world around us. Since its beginnings, Slow Food has grown into a global movement involving **millions of people in over 160 countries,** working to ensure everyone has access to good, clean and fair food. According to Slow Food International Philosophy, food is tied to many other aspects of life, including culture, politics, agriculture, and the environment. Through food choices, people can collectively influence how food is cultivated, produced and distributed, and change the world as a result (Slow Food International).

Europe's anti-GMO movement GMWatch is another example of movements fighting for the interests of consumers. In 1999, European NGOs and concerned scientists met in Brussels to discuss how to prevent an uncontrolled and contaminating of fields and seeds in Europe with genetically engineered organisms. Inspired by the nuclear-free zone movement, the idea to spread GMO-free zones and regions, declared by citizens and their local and regional governments, started to sprout. Since then, the GMO-Free-Regions movement continued to grow and what had started as an European movement spread beyond the continent (GMO-free Europe). In 2020 GMWatch movement is acting globally and provides the public with the latest news and comment on genetically modified foods and crops and their associated pesticides (GMWatch). So even these few examples show that food and consumers' movements can be very different, but they are proponents of healthier and better living.

7.2.2 Case Study 'Slow Food Alytus'

1st stage: Emergence of social movement 'Slow Food Alytus'. The emergence of the club 'Slow Food Alytus' was initiated by the international organization 'Slow Food' when the representatives of this organization got in contact with Džiuginta Rasiukevičienė. She brought together the enthusiasts of healthy food and healthy lifestyle, representatives of the associations of eco-communities, producers of natural agriculture and local products and all together they established the Slow food *convivia* (*convivere* in Latin) in Dzūkija and Dž. Rasiukevičienė was elected its president. Movement 'Slow Food Alytus' was established in 2017.

The goal of the community is to promote the consumption of local, properly grown and produced products, to spread the word about the benefits of good food and to remind everyone to enjoy eating. Slow Food envisions a world in which all people can access and enjoy food that is good for them, good for those who grow it and good for the planet. Their approach is based on a concept of food that is defined by three interconnected principles: good, clean and fair.

GOOD:	quality, flavoursome and healthy food
CLEAN:	production that does not harm the environment
FAIR:	accessible prices for consumers and fair conditions and pay for producers (Slow Food International).

Movement 'Slow Food' cooperates with other local initiatives—'Kaimas veža' ('Countryside Brings'), Healthy Club 'Determination', association 'Native Land Home'. Cooperation between these organizations brings together likely minded people and creates infrastructure between farmers and consumers. The main goal of the 'Slow Food Alytus' is to bring new quality of food and to know how the food was grown, to know the farmer and his/her philosophy. In other words movement 'Slow Food Alytus' is changing the relationships between the farmer, the customer and the food. The goal of the movement had not been changed since its establishment and it still helps customers and farmers to reach each other and to enjoy mutually beneficial cooperation. Džiuginta Rasiukevičienė also emphasized that their philosophy is based not only on slow food, but on a slow living as a such. That is why one of the key aspects of their community is to eat food which is grown not further than 30 km from centre of their *convivia* which is based in Meteliai not in Alytus.

2nd stage: Coalescence of social movement 'Slow Food Alytus'. The need to gather like-minded people and spread the ideas was a natural further step in reaching the main goal (to promote the consumption of local, properly grown and produced products, to spread the word about the benefits of good food and to remind everyone that they should enjoy eating) as it is not possible to establish *convivia* without farmers, their products and consumers. There are two types of community members of the movement 'Slow Food Alytus'—farmers or providers and consumers or receivers and their involvement is hugely different. The first part of members is farmers and there are 12 members of 'Slow Food Alytus' Board. Board members are permanent, and they are producers of natural agriculture and local products, as well as they are small farmers. The number of Board members had not have changed since the establishment of community in 2017.

According to the philosophy of Slow Food International, consumers or eaters also are members of the *convivia* but first of all their involvement is much weaker and second, their number is constantly floating, as new members join *convivia* and some of the old eaters find something different where they want to direct their time and money or they simply change their residence.

However, there can be drawn some common characteristics of 'Slow Food Alytus' *convivia* members—the vast majority of them are people living in rural areas, who are small, eco-friendly, organic, clean and fair farmers, older than 35 years and mostly women. This sociological description of members of 'Slow Food Alytus' *convivia* mostly represents farmers/providers. Consumers/eaters/receivers can be described as people living in more urban areas (but not only), older than 35 years and mostly women. Also, there is a noticeable trend in growing numbers of consumers but as Džiuginta Rasiukevičienė confirmed that higher growth is welcomed and that is something that all of members of *convivia* are working hard on but they do not forget to live sustainable and slow.

However, analysing the development of 'Slow Food Alytus', it must be said, that an expansion or broader scope of the activities is not the main goal of this organization. The President of 'Slow Food Alytus' Džiuginta Rasiukevičienė confirmed that the main idea of their movement is high quality, clean, fair, and slow food and living. Because of this reason, community pays attention to maintaining its members, deepening their cooperation and looking for the new forms of cooperation. For example, 'Slow Food Alytus' organizes food tastings for various companies as a certain leisure activity. Club organizes healthy food tasting for those who, due to their constant busyness, seem to be forced to consume fast food instead of 'slow', organic food.

An important part of the community development is its activities for the members of *convivia*. Club 'Slow Food Alytus' also organizes food tastings for its members, promotes local farmers, prepares food, and uses that to strengthen the sense of the community, togetherness and belonging.

Another step in the development of the organization is its cooperation with other like-minded organizations, such as 'Kaimas veža' ('Countryside Brings'), Healthy Club 'Determination', association 'Native Land Home'. Healthy Club 'Determination' helps in organizing regular meetings and tastings with *convivia* members, 'Kaimas veža' ('Countryside Brings') is an online platform where farmers and eaters meet, and it helps to bring food from the farm to the customer. Finally, 'Slow Food Alytus' closely cooperates with the association 'Native Land Home', which helps oncological patients and together these two organizations help patients in providing healthy food and organizing activities lifting the spirit.

3rd stage: Bureaucratization of social movement 'Slow Food Alytus'. The third cycle stage of any social movement is institutionalization, formalization and bureaucratization where every organization or social movement decides how it organizes its activities, responsibilities and recourses.

Structure of the movement. Organization of 'Slow Food Alytus' has semi-formal structure and to some level informal sharing of activities. Džiuginta Rasiukevičienė is the president of the movement 'Slow Food Alytus', also there is twelve members board, which makes decisions on the most important activities of the movement. Slow food movement is decentralized movement with open membership to join this initiative.

Roles of the movement. Main roles are clearly identified to the members of the movement, for the part of producers: members, responsible for

(1) production; (2) sales; (3) logistics (in cooperation with kaimasvez a.lt platform); (3) tastings; (4) voluntary activities and education. As it was mentioned earlier—eaters are free to decide on the type and form of cooperation between members but their commitment and activities in the *convivia* are scarcer and weaker.

'Slow Food Alytus' can be called as more informal than formal movement or even community, and it offers the following services:

- sales of local food products. Local products such as large variety of seeds, herbs, dairy products, all kinds of vegetables, fruits, berries, rapeseed oil, bread and many more local organic production grown not further than 30 kilometres from eaters and/or supplied to the customers via internet platform kaimasveza.lt.
- food tastings for community members and various companies who want their employees to enjoy slow eating and slow living.
- various voluntary work with healthy clubs, oncological patients and schoolchildren showing them the opportunities to eat healthy and local food.

The main activities of the *convivia* 'Slow Food Alytus' include (1) promotion of the message of good, clean and fair food; (2) promotion of the accessibility to taste and buy good, clean and fair food; (3) helping those who cannot access (because of their age or health) good, clean and healthy food.

Human resources. Human resources such as labour, experience, employees' skills and expertise are the key elements in reaching the goals of social movement. The community of 'Slow Food Alytus' connected healthy living enthusiasts who were active in this field before and even the movement does not have formal staff, but the people who share the same values and ideas created a sustainable environment for like-minded people to reach their goals where producer and consumer meet and share mutually beneficial relationships and create informal network for further cooperation.

Financial resources. Financial resources are necessary at the beginning of any organization or movement; however, the founders of the initiative did not need big financial resources to start their activities and everything they did was a private initiative to reach their goals. Also, movement does not have any membership fee, so any expenses for the development of the movement are paid by board members or earned from *convivia* activities.

Intangible resources. Intangible resources play a major role in the development of any movement. Movement 'Slow Food Alytus' developed successful informal relationships with many other like-minded movements and other actors which helps them to reach out to stakeholders—farmers and consumers, to promote their lifestyle and products, and to share their intellectual, human and organisational capital, knowledge and values about slow food and slow living.

Identification of the movement. Slow food movement is quite well-identified in Lithuania. Members actively spread information on local food and relationships between farmer and consumer in various places: social forums, social media, local newspapers, local and national websites, local TV broadcast and radio, meetings and conferences and discussions. Information about *convivia* 'Slow Food Alytus' can be found at their Facebook group 'Slow Food Alytus'. Movement is identified by promoting local, fresh, good, clean and fair food and promoting the idea of low eating and slow living.

Communication channels. Facebook is the main tool used for spreading information about movement to the wider society; also, it is an internal communication among members of the *convivia*, where they can find information about future events and meetings. Finally, Facebook is a tool to organize work, as many people firstly get in touch with the organization through Facebook—whether they want to order a tasting, to know about the *convivia* or to buy products. Other platforms which are actively incorporated in the activities of 'Slow Food Alytus' are these:

kaimasveza.lt 'Kaimas veža' ('Countryside Brings') is a platform where consumers can order organic and local food products.

Facebook page Healthy Club 'Determination' closely cooperates with 'Slow Food Alytus', as they organize members' weekly meetings, tastings and various education activities about healthy food, slow food and slow living.

All these platforms are used actively, and their cooperation shows great integration between different organizations still having the same goal and sharing the same values—be healthy, stay healthy and help others to do that.

International activities. 'Slow Food Alytus' is a part of international movement 'Slow Food'. Being a part of a worldwide known and acknowledged organization is a great advantage for small and local movements as they can get a lot of great ideas, inspiration and support on how to develop slow food movement in Lithuania.

4th stage: Decline of social movement 'Slow Food Alytus'. Current stage of development of social movement 'Slow Food Alytus' is successful with the prospect for growth in the future. From the beginning in 2017 until 2020 every year new eaters join this movement and participate in tasting events, buy healthy and local food and gradually they start living slow.

When talking about future, Džiuginta Rasiukevičienė assured that even an expansion or broader scope of the activities is not the main goal of this community, but the founders of the movement have strong intentions for bigger growth as they seek to influence local political decision makes and public regarding healthy eating, especially regarding to young schoolchildren. This issue is one of the future goals for the 'Slow Food Alytus' movement. Also, founders of this community believe that for further growth and prosper of their organization additional resources will be needed in near future such as more steady human and material resources in educating society (especially young schoolchildren) on benefits and advantages of healthy and local food and eating and living slow.

This social movement has an impact on the society by proposing: firstly, new quality of food and changing farmer, customer and the most important food relationship; secondly, strengthening new food consumption habits and requesting new quality of food; and thirdly, empowering new livelihoods which are changing rural/urban (business) relationship.

Overall, such movements as 'Slow Food Lithuania' are changing the perception of rural areas to towns residents and show us that ties between rural and urban areas are connected and overlapping more than ever before not only in exchanging of goods and services but in exchanging of lifestyles and finding new relationships.

7.3 New Forms of Accessibility to Food for Consumers for Farmer-Customer Relationship Change

7.3.1 Local Food Movements

In the last couple of decades, the local food movement has gone from a fringe movement to a major player in the national food industries. At this time local supermarkets pretty much had a monopoly on the food industry (Roberts, 2017). For example, people might have been able to choose fruits and vegetables over produced products, but there really were not a lot of options for how to get your food. The local food movement was born out of the industrialization of organics (Ikerd, 2017). It has emerged from the erosion of public trust and confidence in organic foods and aims to bring farmers and consumers closer together in various innovative ways.

The organic food movement emerged as a consequence of declining trust and confidence in the conventional/industrial food system (Ikerd, 2017). Almost that every major urban center around the country has several farmers' markets, community-supported agricultural programmes and other innovative ways to bring farmers and consumers closer together. The local food movement improves access to healthy, organic food, strengthens the local economy, and improves community relationships (Roberts, 2017). Many factors have contributed to the growing popularity of local foods. However, the modern local food movement was born out of the industrialization of organics (Ikerd, 2017).

Most retailers, however, consider local to be on a smaller scale than the state level and the term is often defined as products produced and sold within county lines. There are two primary forms of 'local' when it comes to food: direct-to-consumer (farmer to you) and direct-to-retail/foodservice (farmer to restaurants, hospitals, schools and organizations). Local food is the better choice if attempting to purchase and consume goods in or near your geographic location (Brain, 2012).

Demand for local food has been growing exponentially and is now reaching wholesale and institutional markets. This development is considered a potential step towards solving the above problems but has introduced another concern regarding scale: how can we bring small-scale, sustainable, local food to larger markets, given that scale has historically been inversely related to both sustainability and socio-economic justice? The local food movement entails but is not reducible only to concerns such as production methods and food miles, health and food security or economic and community development (Furman & Papavasiliou, 2018) (Table 7.2).

7.3.2 Case Study 'Local Food Movement by Viva Sol'

1st stage: Emergence of social movement 'Local food movement by Viva Sol'. The main initiator for creation of local food movement from

Table 7.2 Main reasons to consume local food

Main reasons to consume local food
Economic
Farmers retain a greater portion of the value-added costs typically captured by large firms
("middlemen") further down the supply chain. This helps to sustain rural communities and preserve small farms
Purchase of local food ensures more income to local community
Small business are the largest employer
Consume of local products increase nation's food security
Environmental
Eating locally helps to preserve local and small-scale farmland
Reducing the distance food travels cuts down on associated fossil fuel consumption, air pollution, and
greenhouse gas emissions
Supporting local food helps preserve cultivar genetic diversity
Farmers who engage in direct marketing are more likely to use environmentally friendly production
practices
Mental and physical
Local food systems are linked to reduced food safety
risks through production decentralization
Eating locally is correlated with improved nutrition, increased likelihood of making healthier food
choices, obesity prevention, and reduced risk of diet-related chronic
disease. This is mainly because the food is more nutritious, fresher, and less processed
Social:
Gaining of insight into your food's story through talking with the people who grew and/or made it
The ability to talk with producers when purchasing food allows you to ask questions about pesticides,
herbicides, growth hormones, animal treatment, fertilizers, and any other queries you may have about
how your food was produced
Getting to know your local producers gives you a stronger sense of place, relationships, trust, and pride within your community

Source Brain (2012)

cheese products in Lithuania is Valdas Kavaliauskas, who decided to move from Vilnius to Darguziai Village in 2008 to raise goats and to produce goat cheeses. He was looking forward to collaboration with supporters of the idea in this region that would strengthen cooperation ties between the rural and urban population and support ideas of local food movement, and would offer outstanding dairy products to urban citizens. Valdas

Kavaliauskas and other cheese producers from this region who joined this activity founded Viva Sol, which means 'long live the earth' in French, an association to address relations between producers and consumers and between urban and rural areas. In 2008–2009 they decided to set up a Cheese Farm Market in one of the cafés of the capital city and a Cheese House in Darguziai Village. Every year more and more activities were proposed as (1) tasting fest aiming to create strong relation with consumers, (2) collaboration with restaurants in different cities and towns, (3) pick-up points of cheese production in 8 cities and towns (in food market, restaurants, small food shops), (4) cheese school in Darguziai village, (5) production of supplements to produce cheese. Later Valdas Kavaliauskas has moved to another village where he continues his work together with other members of the association. Later association Viva Sol has started to focus more on broad topics related to development of rural areas, sustainability of rural communities, topics with climate change, environment, etc.

Motivation to create local food movement from dairy products was based on the need for high-quality and delicious food. Conventional food system does not allow satisfying your needs for fresh locally produced food. This was one of the most important aspects that led to the initiator decision to create a movement of local food, to move to the countryside and to turn to farming. On top of that, the initiators put forward another important factor—the urge to be close to the nature and to enjoy the feeling of freedom. Such needs can be effectively met by the pattern of their life in the countryside where they practise small farming. Valdas Kavaliauskas, the initiator and pioneer of cheesemakers' activities, says that his participation in Darguziai Village community action and his experience to a certain degree empowered him, acted as a driving force, and encouraged him to develop his activities and to bring like-minded people together. The other two important aspects are changes in the needs of Lithuanian food consumers and foreign experience. They strengthened confidence that such a pattern of relationship between the producers and the consumers may exist.

The main goal of local food movement by Viva Sol is to improve liveliness through rural areas and to create new form of accessibility to food for consumers changing farmers and consumer relationships. Association assists the survival and establishment of small farms by inviting urban citizens to come to the village and to join the activities organized in rural areas, their festivals and farm work. Furthermore, rural people are encouraged to visit the city, to learn about the city life, the activities of urban people and to understand what people distanced from the natural environment think. These actions create a close link between rural and urban areas. Different tools used in pursuance of this objective include various events in rural settings, lectures, discussions and farmers' meetings with urban consumers when they sell their products in the urban environment.

The underlying concept of the association, which has already become a propelling slogan, is 'May the Rural Areas be Alive'! The members of the association describe their philosophy in five sentences:

1. Relationship between the consumer (an urban citizen) and producer/grower (a rural citizen).
2. Solidarity between all members of the association and supporters of the association members' ideas.
3. Certainty and stability, i.e. the consumer should know how and where to obtain a real home-grown and handmade product.
4. Trust among consumers and producers. The members of the association believe that it is highly important for the producer to know the consumer and vice versa as this gives birth to something highly important, which is trust.
5. And therefore the association is always ready to talk about its activities.

The main goal remained unchanged from the establishment of the association until 2020. Association Viva Sol has 18 members, including 2 organizations (Baltic Environment Forum, Vocational School 'Garden masters'), 5 farmers and 11 end consumers.

2nd stage: Coalescence of social movement 'Local food movement by Viva Sol'. The increase of the members of the local food movement by Viva Sol was very natural and based on activities proposed by the association. Producers—one part of members of association—were willing to cooperate on the distribution of production from cheese with the involvement of consumers aiming to create strong relationships with producers. Strong relations with consumers are created by various events and initiatives organized by initiators of the association Viva Sol. For example, the Cheese House in Darguziai village organizes celebrations of public holidays, community volunteering, open door days, discussions, encourages sharing of farming and life experience, and invites volunteers, who are willing to learn about the farming and cheese-making process.

Another reason helping to expand number of members in this movement is the origin of this idea. The birth, development and implementation of this idea were set in motion by a similar worldview of the participants and examples of comparable practices in other countries. Here the initiators of the idea relied on the French practice. You could say that the Lithuanians did not only bring home French cheese-making secrets but also adopted the idea of solidarity and openness. They long to see rural areas alive with their crafts, businesses and agriculture. Moreover, the cheesemakers believe that rural viability highly depends on the relationship between rural and urban citizens. Association Viva Sol was founded by both rural and urban people, who cherish similar values. Initially they were encouraging others to develop relationship between urban and rural citizens.

Members of movements are producers and consumers. The part of producers includes small farmers from rural areas, who work towards producing high-quality and tasty food.

The Cheese House in Darguziai village has quite a considerable number of partners that can be broken down into 3 groups:

1. Small partnership of Cheese Experts engaged in cheese wholesale and retail and organization of events and tasting.
2. Cooperative 'Our Cheeses'.
3. Farmers cheesemakers.

The group of consumers embraces mostly urban citizens with various professions who appreciate and look for high-quality and tasty food. *The target group of users includes urban citizens. Consumers of cheese products are medium aged from 35 to 55 years.*

Every year is continuous growth of new members of consumers who firstly participate in cheese fest to try various species of cheeses and then continue to consume it. Remaining part of members—producers—is stable with no changes from establishment of the association.

3rd stage: Biureaucratization of social movement 'Local food movement by Viva Sol'. *Resources of the movement.* Two organizations were set up for the purpose of local food movement development: 5 small farmers, members of association Viva Sol, are manufacturer, and an association Viva Sol seeks to establish relationship between urban and rural areas as well as to improve liveliness through rural areas. The main initiator of both organizations is the originator of the idea Valdas Kavaliauskas, who invited other like-minded persons to join the process.

The members describe their organization as a live farm of farmers cheesemakers, which offers the following services:

sales of local food products. Local products as large variety of cheeses, bread, cream, caramel and other delights produced by the cheesemakers are sold in Cheese House in Darguziai or supplied to the pick-up points in 8 cities or small towns to consumers from cities.

catering: you can order different locally made dishes (e.g. a hot vegetable soup, a Cheesemaker's casserole, salad from home-grown vegetables, home-made ice-cream for dessert and refreshing kvass, which is a drink made of bread).

Other services provided by the Cheese House include:

- cheese tasting;
- demonstration of the fermented cheese manufacturing process to groups of visitors;
- a sight-seeing tour around Darguziai Village;
- various voluntary work on the farm (e.g. goat herding and other farm work).

The main activities of the association Viva Sol include (1) promotion of the survival and establishment of small farms by inviting urban citizens to come to the village and to join the activities organized in rural areas, their festivals and farm work and (2) encouraging rural citizens to visit the city, to learn about urban life and activities of urban population.

Human resources. Human resources represent the key element in the development of the local product market. It is extremely important to have an idea, initiative, like-minded people and willingness to create and to make one's contribution to building sustainable environment by using other essential elements. In the case of this good practise, human resources were the keystone in the successful implementation and further development of this initiative, attracting new like-minded people and a large group of consumers, i.e. creating an informal network.

Financial resources. Financial resources are necessary at the beginning for buying land, livestock and machines for production, unless there is initial capital (e.g. land, equipment, farm animals). According to the originators of the initiative, they did not need big financial resources to start their activities as in the rural areas of Lithuania land and other inventories are not expensive and thus it is not necessary to take big credits or to assume other liabilities that would render this activity less attractive.

Intangible resources. Intangible resources, such as the establishment of informal relations among the producers and between producers and consumers, innovative marketing campaigns, promotion of direct sales, selling to people visiting the farm, consumer involvement in different activities, events, etc., play a major role in seeking successful operations.

Structure of the movement. Association Viva Sol is formal organization. Local food movement is informal decentralized movement with open membership to join this initiative.

Roles of the movement. Main roles are clearly identified to the members of the movement, for the part of producers: members, responsible for (1) production; (2) pre-order; (3) logistics; (3) workshops at Cheese School; (4) tasting fests; (5) activities in Cheese maker house. Consumers are free to decide on the type and form of cooperation between members.

Identification of the movement. Local food movement by Viva sol and Cheese house is very well-identified in Lithuania. Members actively spread information on local food and relationships between farmer and consumer in various places: social forums, TV broadcast, radio, newspapers, meetings and conferences, discussions organized by various public authorities, etc. All information can be found at their platform www.surininkunamai.lt. Movement is identified by promoting local food viability, mobility and establishment of relations.

Communication channels. Platforms www.surininkunamai.lt and Facebook are tools used for (1) internal communication among members of the network; (2) spread information about movement to the wider society; (3) tool to organize work. Both platforms are used very actively. Initiators of movement are responsible for maintaining work of both platforms.

Viva Sol also actively participates in the international projects as a main partner or project partner (for example, Erasmus plus programme). This enables them to spread their experience with partners from abroad and also to bring fresh new ideas to the movement.

4thth stage: Decline of social movement 'Local food movement by Viva Sol'. Current stage of development of social movement 'Local food movement by Viva Sol' is successful with prospect for some growth in the future. From the beginning in 2008 until 2020 every year new consumers join this movement so-called 'cheese lovers' that firstly participate in cheese tasting fest, events and finally choosing their own way how to get food basket of cheese every week. Initiators of the movement have no intention for bigger growth and they let movement develop in natural way. If there would be a need for more members in this movement, additional resources would be needed as (1) education of society on benefits of short supply chain; (2) advantages of fresh local food; (3) more distribution tools/channels/members-producers, etc.

This social movement have an impact on the society by proposing (1) new form of accessibility to food for consumers changing farmer and customer relationships; (2) advocating for new small-scale family farming, (3) helps to get a better price both for farmers and consumers; (4) creating and/or strengthening new food consumption habits and requesting new quality of food.

7.4 New Livelihoods for Rural/Urban Businesses Relationships Change

7.4.1 *Back-to-the-Land International Movements*

Movements 'back-to-the-land' are one of the most interesting and most recent movements in the post-industrial movements' era. And even in the Western part of the world, 'back-to-the-land' social movements became greatly popular in the post-Woodstock era in the late 1960s and early 1970s, supported by a mostly idealistic group of people who wanted to live life more simply (Jeffrey & Merlin, 1986); in Eastern part of Europe, this movement was late for at least fifty years and only in 2000s and 2010s social movements 'back-to-the-land' arose and are still trying to get the public attention. There are many international and regional back-to-the-land movements across the world and namely, the most known movement is La Via Campesina International Peasants' Movement (LVC), as well as others movements inspired by La Via Campesina, such as 'Reclaim the Fields' and many others national back-to-the-land movements.

The pioneer movement La Via Campesina is a global, transnational movement acting in the field of food sovereignty, climate and environmental justice, international solidarity, agroecology and peasants' seeds, peasants' rights, land, water and territories, dignity for migrants and waged workers. In 2020, La Via Campesina unites 182 organizations,

81 countries and more than 200 million peasants (LVC). LVC was established in 1993 and it is a global alliance of organizations of family farmers, peasant farmers, indigenous people, landless peasants, farm workers, rural women and rural youth, representing at least 200 million families worldwide. In Europe, La Via Campesina unites 26 national organizations from 17 countries.

The establishment of the movement 'Reclaim the Fields' was inspired by the La Via Campesina and the movement is promoting food sovereignty and peasant agriculture, particularly among young people and urban dwellers, as well as alternative ways of life. The movement is seeking to create alternatives to capitalism through cooperative, collective, autonomous, real-needs-oriented, small-scale production and initiatives. By linking local practical action with global political struggles. Movement 'Reclaim the Fields' participates in local actions through activist groups and cooperates with existing initiatives and it is not a homogeneous group but opens up to the diversity of actors fighting the capitalist food production model. Movement address the issues of access to land, collective farming, seed rights and seed exchange (Reclaim the Fields).

7.4.2 Case Study 'I Choose Countryside - Settlers and Similar Hipsters'

1st stage: Emergence of social movement 'I choose countryside - settlers and similar hipsters'. In Lithuania, there are at least four social movements, which can be called as 'back-to-the-land', but their activities and the level of formalization, organization and bureaucratization are still in the start-up phase. One of these movements is 'I choose countryside - settlers and similar hipsters', which was established only in 2019 as a Facebook group uniting like-minded people. The founder of this group is Kotryna Meidė. The idea to implement this initiative arose with the fundamental change in the lifestyle—i.e. moving from the city to a remote village when the leisure and the circle of friends have changed significantly (more precisely, it has decreased significantly). Therefore, she took care to find like-minded people, because it was clear from various articles on the internet that they (K. Meidė and her husband) were not alone. Kotryna Meidė and her husband have family farm where they grow asparagus and renovate their old house, because as Kotryna says and as it is written in their Facebook group—'Here we choose a nature-friendly village with all its big and small pluses. It is more interesting for us to revive an old homestead than to build A+++++. We try to create, grow, share... Because there is more air in the village'.

This initiative largely addressed the lack of support from others. What they wanted the most in creating this community was the support to each other and to show the encouragement to those who are just thinking about settling in the countryside. It is a big step and absolutely not following the crowd. In Kotryna Meidė opinion, hesitation can keep many people away from the dream of living in the countryside. That is why finding like-minded people is essential in seeking this dream.

The main goal of the movement 'I choose countryside - settlers and similar hipsters' is to bring together like-minded people, and it had not changed since the establishment of the movement. The founder of the movement 'I choose countryside - settlers and similar hipsters' also confirmed that it is a pity that there is a lack of time to incite topics and encourage members and settlers to share their stories.

To sum up the first stage of life cycle—emergence—of social movement 'I choose countryside - settlers and similar hipsters' it has to be said that the movement has a good start, it cleared its ideas and values, which shows that this movement empowers new livelihoods changing rural/urban relationships, brings new forms of rural lifestyle by changing relationship with neighbours—where neighbours can be not in their physical place but in online as well.

2nd stage: Coalescence of social movement 'I choose countryside - settlers and similar hipsters'. The main reason for deciding to gather at least an online community of like-minded people and spread the common ideas more widely was the difficulties of settling in the village at the beginning. First of all, for many people, it is difficult to decide to go to live in the village as there was and there are a lot of hostile attitudes, lack of confidence and enthusiasm, discouragement. These are serious challenges for many people. Founders of 'I choose countryside - settlers and similar hipsters' strengthened themselves and realized that they were not really wrong and finally they decided that someone might lack the support they needed in the beginning. As Kotryna Meidė said—'Maybe it will be the straw that a person will grab when traveling to their dream. And what is that dream - a countryside - good for? Everything! Movement, health, freedom, cheap living, freedom for creativity and business and SLOW pace of life'. So, the main reason to create a community 'I choose countryside - settlers and similar hipsters' was founders' experience and need for support in moving to the countryside.

In 2020 community unites about 1700 members but not all of them are highly active. Members get involved not as actively as the founders would like. However, there are real settlers who share their extraordinary adventures in the countryside, the results of their work, seek for advice and give advice, looking for like-minded people. Members who are involved in the movement are mostly young and active people, who seek to be hosts of their lives. Advertising the movement and its members' activities are on the rise, so the growth of the community is expected. Also, it cannot be said that there is a huge change in members number—the number is slowly growing but as the group is informal it is difficult to draw any more detailed conclusions about characteristics of the members of the community. However, there are some members who share their experience, ideas or seek advice more actively and it shows that the group implements the main goal of itself—to support each other when difficulties arise.

3rd stage: Bureaucratization of social movement 'I choose countryside - settlers and similar hipsters'. *Structure of the movement.* Structure of the movement 'I choose countryside - settlers and similar hipsters' is informal and decentralized. Members communicate via Facebook group or individually. Organization of the movement is based on voluntarily basis and the head of the movement as well as the only one administrator of the Facebook group 'I choose countryside - settlers and similar hipsters' is Kotryna Meidė. The membership of the community is informal and open to any person who supports ideas of the movement and is looking for some advises or willing to share their personal experiences.

Roles of the movement. Roles of the movement are informal. Kotryna Meidė administers a Facebook group where all members can share their information, invite members of the group to various events and meetings online or in person.

Resources of the movement. The most important resources of the movement are human and intangible resources. The movement 'I choose countryside—settlers and similar hipsters' connected countryside enthusiasts who were active in this field before and even the movement does not have formal staff, but the people who share the same values and ideas created a community for like-minded people to reach their goals where they can share their experience, challenges and success stories. Intangible resources play a major role, as the movement 'I choose countryside - settlers and similar hipsters' is still developing successful informal relationships with many people which helps them to promote their lifestyle and to share their intellectual, human and organizational capital, knowledge and values about living tin the countryside. The community does not have any membership fee or formal financial resources.

Identification of the movement. In the last five years movements 'back-to-the-land' are becoming more popular in Lithuania and members of these movements are the best ambassadors in spreading their ideas and values. Movement 'I choose countryside - settlers and similar hipsters' spreads information about living in countryside by using various channels of information, such as social forums, social media, local newspapers, local and national websites, local and national TV broadcast and radio and informal meetings. At this stage of the development of the movement, Kotryna Meidė is doing the biggest part of these activities as there is still the lack of involvement of other members. In the future, the founders of the movement would like to devote more time to promoting the settlement in the countryside and everyday stories or at least to find someone to help them to manage the activities of the group. Community does not have any traditional events yet but there can be found information about various events organized by members of the movement in their Facebook group where everyone is invited.

Communication channels. The Facebook group serves as a platform for internal communication among members of the community also it is a tool to spread information about the movement to the wider society. Many members of the community, as well as the founder Kotryna Meidė, use their private Facebook accounts and/or their agricultural business's accounts to promote living in the countryside. Movement 'I choose countryside - settlers and similar hipsters' is not a member of any international organization yet.

4th stage: Decline of social movement 'I choose countryside - settlers and similar hipsters'. It is difficult to evaluate the fourth life cycle of this movement as it was established only one year ago—i.e. in 2019. The current stage of development of social movement 'I choose countryside - settlers and similar hipsters' is quite successful with continuous growth for their first year of activity. The founder of the movement expressed her concerns that the activity of the movement is quite vague at the moment. However, changing relationships between urban and rural areas and young enthusiasts of the countryside have all possibilities to expand their activities, to promote living in countryside and help each other at the same time. Timing is favourable for the movements 'back-to-the-land' and as this case study and the case analysing ecovillages show—Covid-19 revealed the countryside has to offer much more than cities in the time of crises.

Founders of the movement would like the greater growth of the movement in reaching out for more members and in deepening the cooperation between members, organizing more activities together, uniting members of the movement and sharing each other's experiences and success stories. Also, founders of this community believe that for further growth and prosper of their organization additional resources will be needed in near future such as human and material resources to reach more persons interested in living in the countryside.

Kotryna Meidė is quite precocious in evaluating the impact of the movement 'I choose countryside - settlers and similar hipsters'. She assured that she and her husband have heard that like-minded people decide to move to the countryside (or started dreaming about it more boldly) following their personal story on moving to countryside and starting their asparagus farm (meidukis.lt). The influence of the group has not yet been heard. However, from the reactions to the shared stories, it is quite clear that people are interested in the life of rural settlers and people are looking for ideas and courage. Looking from more scientific perspective it is clear that social movement 'I choose countryside - settlers and similar hipsters' has an impact on the society by proposing: firstly, empowering new livelihoods which are changing rural/urban (business) relationship; secondly, it helps to reorganize rural community life; thirdly, it creates new forms of rural lifestyle-changing relationships with neighbours where neighbours are not only next to each other physically but also the creation of 'online neighbours' where you can get some help, support or advise.

7.5 New Forms of Rural Lifestyle for Changing Relationship with Neighbors

7.5.1 Ecovillages Movements

Creation of global ecovillages movement goes back to the last decade in the last century and counts about 30 years of experience. The impulse for the Global Ecovillage Network (GEN) evolved from an initiative taken by Gaia Trust in 1990 where Ross and Hildur Jackson from Denmark were the driving forces in creating the GEN network. In 1991, the

Gaia Trust convened a meeting in Denmark of representatives of eco-communities to discuss strategies for further developing the ecovillage concept. That led to the formation of the Global Ecovillage Network. In 1994, the Ecovillage Information Service was launched. In 1995, the first international conference of ecovillage members, entitled Ecovillages and Sustainable Communities for the twenty-first century, was held at Findhorn, Scotland. The movement grew rapidly following this conference. Currently, the network is made up of approximately 10,000 communities and related projects where people are living together in greater ecological harmony (GEN History, 2015; Global Ecovillage Network, 2020a). Network members share ideas and information, transfer technologies and develop cultural and educational exchanges.

The Global Ecovillage Network (GEN) catalyzes communities for a regenerative world. GEN is a growing network of regenerative communities and initiatives that bridge cultures, countries and continents. GEN builds bridges between policymakers, governments, NGOs, academics, entrepreneurs, activists, community networks and ecologically minded individuals across the globe in order to develop strategies for a global transition to resilient communities and cultures. GEN is composed of 5 regional networks: (1) Latin America (CASA), Council of Sustainable Settlements of Latin America; (2) North America (GENNA); (3) Africa (GEN Africa); (4) Europe (GEN Europe); (5) Oceania Asia (GENOA), Gobal Ecovillage Network Oceania & Asia. The fifth network also was created in October 2005 with a focus on youth arm, NextGEN, spanning the globe (Global Ecovillage Network, 2020b).

Goals of GEN. The following goals are identified by the network:

- To advance the education of individuals from all walks of life by sharing the experience and best practices gained from the networks of ecovillages and sustainable communities worldwide.
- To advance human rights, conflict resolution and reconciliation by empowering local communities to interact globally, while promoting a culture of mutual acceptance and respect, effective communications and cross-cultural outreach.
- To advance environmental protection globally by serving as a think tank, incubator, international partner organization, and catalyst for projects that expedite the shift to sustainable and resilient lifestyles.

- To advance citizen and community participation in local decision-making, influencing policymakers, and educating the public, to accelerate the transition to sustainable living.

Definition of Ecovilagges by GEN. An ecovillage is an intentional, traditional or urban community that is consciously designing its pathway through locally owned, participatory processes, and aiming to address the ecovillage principles in the 4 areas of regeneration (social, culture, ecology, economy into a whole systems design). Ecovillages are living laboratories pioneering beautiful alternatives and innovative solutions. They are rural or urban settlements with vibrant social structures, vastly diverse, yet united in their actions towards low-impact, high-quality lifestyles. (Global Ecovillage Network, 2020b). Ecovillages provide models for a lifestyle that reduces ecological footprint while delivering a better quality of life: one, which is possible in all countries of the world, and can lead to global justice, solidarity and cooperation. Ecovillages are aiming to learn how to solve conflicts how to develop a global consciousness, how to create places where children can grow up naturally, how to use renewable integrated energy systems, 100% of organic food productions and how to live lives full of love and compassion (GEN History, 2015).

Some examples of ecovillages, network members:

- Sarvodaya (2,000 active sustainable villages in Sri Lanka);
- the Federation of Damanhur in Italy;
- REDES in Senegal;
- small rural ecovillages like Gaia Asociación in Argentina and Huehuecoyotl in Mexico;
- Network also includes urban rejuvenation projects;
- Los Angeles EcoVillage;
- Christiania in Copenhagen;
- permaculture design sites such as Crystal Waters, Australia, Cochabamba, Bolivia and Barus, Brazil;
- educational centres such as Findhorn in Scotland, Centre for Alternative Technology in Wales, Earthlands in Massachusetts, and many more.

7.5.2 Case Study 'Lithuanian Network of Ecovillages and 'Kin's Domain' Settlements'

1st stage: Emergence of social movement 'Lithuanian network of ecovillages and 'Kin's Domain' settlements'. Initiators of Lithuanian network of ecovillages and 'Kin's Domain' settlements were inspired by the ideas in the series of books 'Ringing Cedars of Russia' by Vladimir Megre and his heroine Anastasia. First meetings of initiators for discussion on their vision have started in 2002. Known as 'Kin's Domain' settlements, they are often called new generation of ecovillages with a focus on sustainability for future generations of families (see more at Kin's Domain Concept, 2020; Megre, 2020). First settlements based on this 'kin's domain' concept were established in 2001–2002 in different post-soviet countries (Russia, Lithuania, Latvia, Belarus, etc.). Idea proposed in the book was attractive because it proposed individual living instead of completely collective living and this was more appealing in these countries because of the experience of forced collectivization during the Soviet era (Vidickiene, 2013). Some years later, Vladimir Megre's books have been translated into other languages and these types of settlements have become popular not only in Europe but also in the whole world. Ecovillages based on the 'kin's domain' concept are developed as settlements consisting of individual homesteads. Each family or individual has at least 1 hectare of land, developed as a self-sustainable ecosystem. One hectare of land is large enough to allow a complete, closed cycle of energy and matter. The boundary of a 'kin's domain' is a living fence consisting of trees, bushes and shrubs. It ensures protection from the wind, gives shelter to wild animals and serves as a natural demarcation of the person's realm of the family. One-quarter to three-quarters of the area is covered with perennials: forest and fruit trees and bushes. Rich native species are planted on the principles of enhancing biodiversity. Hundreds of plant species are linked by the principles of permaculture to facilitate the coexistence of the different plants. Recycling of natural materials provides plants with nutrients and increases the quality and quantity of the crop. If no natural water resource is in the territory of settlement, a small pond is dug. Houses are constructed from natural materials. Food is produced using permaculture principles: without ploughing, weeding, pruning, fertilizing or spraying with chemicals. The inhabitants of the ecovillage preserve their natural heritage, taking into account cultural traditions and the ancient knowledge of our ancestors. The idea of a 'kin's domain' is also an idea of a 'little homeland' for many generations. The domain should be a 'space of love' where men, women, children, plants, trees, birds and animals should live in love and care for each other (Vidickiene, 2013).

First settlements in Lithuania have been created in 2006–2007. The peak was reached in 2008–2010. Until 2020 there is continuous growth of 'kin's domain' settlements in the whole territory of Lithuania every year. The biggest motivation for many initiators from Lithuania to join this movement was will to have good environment to grow children, to have fresh air, clean water and good quality food. Concept of 'kin's domain' had all elements of their wishes. Initiators have started to purchase land and to start establishing a settlement in their way by experimenting and discussing experiences with other members in annual meetings. In 2020, there are 6 ecovillages in Lithuania and 40 'kin's domain' settlements. They all are open to invite new members to create their settlements close by to already existing ecovillages or settlements or to create a new place. The great interest to join this movement has appeared in 2020 at the time of the Covid-19 pandemic situation as more land close to already existing ecovillages was purchased. People have understood the advantages of having their own settlement and being in nature instead of living in cities.

The aim of social movement 'Lithuanian network of ecovillages and 'Kin's Domain' settlements' did not change significantly since the first initiatives in Lithuania. It was only supplemented with the need to find more tools for socialization (with neighbours, with other members of movement, with education infrastructure for children).

The main goal is new forms of rural lifestyle-changing relationships with neighbours. The number of inhabitants of ecovillages and 'Kin's Domain' settlements' in Lithuania in 2020 is approx. 500.

2nd stage: Coalescence of social movement 'Lithuanian network of ecovillages and 'Kin's Domain' settlements'. Social movement of ecovillages and 'Kin's Domains' settlement was increasing continuously from the beginning of the movement in 2002. New potential members were participating in annual meetings of the network or approaching residents of ecovillages individually. Platform of ecovillages www.ekogyvenvietes.lt was also a successful tool for live interaction between members of movement and potential residents who were looking for land to purchase or to get more information on principles that are used by ecovillagers.

There is a big potential for growth of the movement in Lithuania as many inhabitants from cities are more concerned about their health, quality of food and environment. Additional tools should be used to increase knowledge of the lifestyle proposed by this social movement.

Currently, about 500 people actively participate in this movement. Every year movement organizes annual meetings in winter, spring, summer and autumn in different locations in Lithuania that all members would be able to join meetings. The number of participants in every meeting is about 150–200. Most of the inhabitants of ecovillages and 'Kin's Domain' settlements are previous city residents with various professions. One of the initiators Raimundas Vaiciunas highlighted that many of the members have professions that allowed them to work remotely (for example, IT specialists). It is very convenient to live in nature, work remotely and travel to city only for some days per week and/or month. Inhabitants are families, so number of men and women are equal. The first inhabitants were approx. 40 years old, families with children aiming to grow their children in natural environment. Later younger people were joining the network, about 30–35 years old with already some experience and vision for their future why they are choosing this lifestyle. There is a very insignificant change of members of the network. Mostly only new people join this movement.

3rd stage: Bureaucratization of social movement 'Lithuanian network of ecovillages and 'Kin's Domain' settlements'. *Resources of the movement.* The most important resources of the movement are human resources. Inhabitants of ecovillages and 'Kin's Domain' settlement use their own efforts in many events and cases when they are developing any activity. Financial resources are needed when they organize some events or meetings. Financial questions for each new initiative are discussed individually how to get resources or who can support it and how. Movement has no permanent place for the meeting or any other material resources. Resources for communication (as website/platform www.ekogyvenvietes.lt, Facebook platform) are developed voluntarily. Recently, there is no need for some additional resources or lack of some resources.

Structure of the movement. Structure of the movement is informal and decentralized. Members communicate via communication channels used by the network or individually. The organization of each event or occasion is discussed beforehand and based on voluntarily basis. Membership in the movement is open to any person who supports ideas of ecovillages and 'Kin's Domain' concept and wants to purchase land and build their settlement.

Roles of the movement. Roles in the movement are partly devoted. All distribution is based on voluntarily basis. There are responsible persons for communication (maintenance of website/platform www.ekovyvenvietes.lt and Facebook platform), organizing movement events (each event is taking place in different settlements and inhabitants of these settlements are responsible for all organizational issues).

Identification of the movement. Ecovillages and 'Kin's Domain' concept is becoming more popular in Lithuania every year since the establishment of the movement in 2002. Members of the network are the main actors that spread information about ecovillages and 'Kin's Domain' settlements in Lithuania using various channels as platforms, TV broadcast, websites that help to create the identity of the network. Movement covers the whole area of Lithuania as inhabitants of settlements are from different places from Lithuania in rural areas and places close to cities. Traditional events 4 times a year (winter, spring, summer and autumn) are organized every year as annual meetings involving about 150–200 participants each time.

Communication channels. Website/platform www.ekogyvenvietes.lt and Facebook platform are tools used for (1) internal communication among members of the network; (2) spread information about movement to the wider society; (3) tool to organize work. Both platforms are used very actively. Initiators of movement are responsible for maintaining the work of both platforms.

Some members of the network also participate in some other events internationally and have good contacts with members from ecovillages in other countries. Movement itself is not a member of any international organization.

4th stage: Decline of social movement 'Lithuanian network of ecovillages and 'Kin's Domain' settlements'. Current stage of development of social movement 'Lithuanian network of ecovillages and 'Kin's Domain' settlements' is successful with continuous growth from the beginning until 2020. Covid-19 pandemic situation has opened even more potential for the movement as many city residents have expressed their interest to live in rural areas and being more independent in nature. Pandemic situation also has proved that many works can be done remotely so there is no need to spend all time in offices. A lot of free land around ecovillages or 'Kin's Domain' settlement was purchased in 2020 at the time of restrictions of Covid-19 pandemic situation.

From the beginning of the movement in 2002 until 2020 every year new enthusiasts join the network. Comparing the period from 2014 to 2020, increase in ecovillages and 'Kin's Domain' settlements doubled, from 19 ecovillages and settlements in 2014 to 44 ecovillages and settlements in 2020. Initiators of the movement have no intention for bigger growth and they let movement develop in natural way. If there would be a need for more participants in this movement, additional resources would be needed as (1) education of society on nature-friendly ways to live with nature, by developing new ways of living in nature-friendly and healthy way; (2) new form of rural lifestyle-changing relationships with neighbours; (3) new programmes in universities and in schools.

This social movement has an impact on the society by proposing: (1) to reorganize rural community life; (2) new form of rural lifestyle-changing relationships with neighbours; (3) advocating of new type of living in rural areas with potential of freeing us of being so dependent on the individualistic, consumerist and commodified system in the cities. They have the potential of letting us learn how to live together on the land again in a genuinely more sustainable way, which is especially important given the climate crisis and resource shortages that we face. Ecovillages are creating low-impact, environmentally harmonious living situations, pioneering nature-friendly agriculture techniques as well as businesses and education centres; (4) creating and/or strengthening new food consumption habits and requesting new quality of food.

References

Brain, R. G. (2012). *The local food movement: Definitions, benefits, and resources*. Logan: Utah State University.

Cheese-Makers House. www.surininkunamai.lt. 3 August 2020.

Food and Agriculture Organization of the United Nations. (2018a). *Second international symposium about agroecology*. http://www.fao.org/about/meetings/second-international-agroecology-symposium/about-the-symposium/en/. 19 November 2020.

Food and Agriculture Organization of the United Nations. (2018b, 3–5 April). *Scaling up agroecology initiative. Transforming food and agricultural systems in support of the SDGs*. A proposal prepared for the international symposium on agroecology. http://www.fao.org/3/I9049EN/i9049en.pdf. 16 November 2020.

Furman, C. A., & Papavasiliou, F. (2018). Scale and affect in the local food movement. *Food, Culture & Society, 21*(2), 180–195. https://doi.org/10.1080/15528014.2018.1427926.

Gelcer, F. J. (2012). *Simbioz c mikroorganizmami - osnova žizni rastenij.*

GEN History. (2015). The history of the global ecovillage network: 1991-2015. https://ecovillage.org/gen-history/. 18 November 2020.

Global Ecovillage Network. (2020a). https://en.wikipedia.org/wiki/Global_Ecovillage_Network. 18 November 2020.

Global Ecovillage Network. (2020b). https://ecovillage.org/about/about-gen/. 18 November 2020.

GMO-free Europe. https://www.gmo-free-regions.org/who-we-are.html. 17 November 2020.

GMWatch. https://www.gmwatch.org/en/. 17 November 2020.

Ikerd, J. (2017). Soul of the local food movement. *Journal of Agriculture, Food systems, and Community Development*, *7*(4), 5–7. https://doi.org/10.5304/jafscd.2017.074.002.

Jeffrey, J. C., Merlin, B. B. (1986). Alternative technology and part-time, semi-subsistence agriculture: A survey from the back-to-the-land movement. *Rural Sociology, 51*(1), 45–59. College Station, TX.

Johnson, M. (2012). *Indigenous agroecology*. http://www.maramatanga.ac.nz/project/indigenous-agroecology. 26 October 2020.

Kin's Domain Concept. (2020). https://kinscommunity.com/kins-domain-history/. 16 June 2020.

Latham, J. (2016, September 20). *Why the food movement is unstoppable. independent science news for food and agriculture*. https://www.independentsciencenews.org/health/why-the-food-movement-is-unstoppable/. 17 November 2020.

LVC. La Via Campesina International Peasant's Movement. https://viacampesina.org/en. 15 November 2020.

Megre, V. A. (2020). *I exist for those for whom I exist.* http://vmegre.com/en. 16 June 2020.

Meidų ūkis. https://meidukis.lt. 15 September 2020.

Platform Kaimas veža. https://www.kaimasveza.lt/. 31 August 2020.

Reclaim the Fields. https://reclaimthefields.org/. 15 November 2020.

Roberts, T. (2017). *The local food movement as a catalyst for community.* https://www.permaculturenews.org/2017/08/22/local-food-movement-catalyst-community/. 19 October 2020.

Roman-Alcala, A. (2018). *Agroecology is becoming a global movement. But where does the U.S. fit in?* https://www.yesmagazine.org/environment/2018/05/07/agroecology-is-becoming-a-global-movement-but-where-does-the-us-fit-in/. 19 November 2020.

Slow Food International. https://www.slowfood.com/. 31 August 2020.

The 2030 Agenda for Sustainable Development. (2015). https://ec.europa.eu/environment/sustainable-development/SDGs/index_en.htm. 26 October 2020.

Vidickiene, D. (Ed.). (2013). *Living in harmony: Inspiring stories from ecovillages*. Vilnius.

Website of the Social Movement of Natural Agriculture in Lithuania: www.gerazemdirbyste.lt. 13 July 2020.

Курдюмов, Н. И. (2012a). *Умный сад и хитрый огород*. Владис, ISBN: 978-5-9567-1611-3.

Курдюмов, Н. И. (2012b). *Мир вместо защиты* (In Lithuanian language: Santarvė vietoje kovos, vert. Saulius Jasionis). ISBN: 9786099531311. Kaunas: Algirdas Katilevicius Publisher.

CHAPTER 8

Conclusions and Discussion: Rural Social Movements in the Context of Rural Development Paradigm Shift

Rita Vilkė and Dalia Vidickienė

Despite the plenty of literature devoted to social change and social movements, which goes back to the last century and keeps being published nowadays, most of them are focused on one specific issue and do not provide theoretical insights on the big picture. At the beginning of the twenty-first century, the belief, that social change is caused or determined by a single factor, which stands for reductionist and determinist theories, was often criticized due to its outdated incapability to go in line with the modern sustainability thought, which undoubtedly has a multifaceted nature. Recent scientific discussions most often agree that social change in current times occurs in the interaction of both groups of factors—systemic and particular unique, which represent the interactionist branch of theories. The interaction of multiple groups of factors in every unit of analysis, e.g. every case of social change, and especially in a form of new social movements, is unique and complex, and thus calls for the particular combination of research attributes to be activated in such area of research, putting the qualitative measures in the first place.

Overall social change, which is broadly understood as a philosophical idea or paradigm advocating sociocultural evolution, which moves the society forward compared to the previous state, enables to examine the

R. Vilkė et al., *Rural Economic Developments and Social Movements*, Palgrave Advances in Bioeconomy: Economics and Policies, https://doi.org/10.1007/978-3-030-71983-8_8

social movements in the context of industrial and post-industrial social movements. However, the conventional distinction between the old and new social movements as representatives of the post-industrial era is often criticized as hardly beneficial for the development of theory.

This monograph is driven by the idea to explore social movements through the prism of paradigm innovations. Such an approach allows providing new insights for the scientific debate between the industrial and post-industrial social movements. The concept of paradigm innovation is broadly related to the new generation of post-industrial movements and may become an umbrella and fruitful direction for future research on all kinds of contemporary social movements, by explaining their difference in character than movements of the past, including goals, stakeholders, strategy and organizational structure.

The shift to post-industrial society is a very complex and wide-scale process that requires many paradigm innovations as radical changes in our mental models and behaviour. Not all of them are well understood. For this reason, this monograph is focused on the examination of social movements dealing with rural development. The social movements for rural development currently cover all so-called 'grand challenges' to society and are the most active power for radical social changes. The literature review and empirical research demonstrate how many aspects cover a new generation of social movements for rural development acting as a paradigm innovations' driving force at the current stage of society's development.

The key area for empirical research has been chosen Lithuania because the countries, experiencing very rapid processes of entering the market economy are the best laboratory for academic research on social movements, especially for industrial ones. Since radical agrarian transformation brought by the industrialization of farming and capitalism's penetration of the countryside help more clearly define the areas of interest and strategies of farmers' movements involved in the industrial agribusinesses system, case studies help deeply illustrate how dramatic these changes occurred in a form of the social movement. Lithuania, like other post-socialist countries, has entered the global agri-food market in the 1990s. This period was generally unfavourable for world food and agriculture. Therefore, social and economic failures of the industrial agri-food system have spurred the active formation of local and national social movements of farmers. The six case studies from Lithuania demonstrate how the national social movements of farmers are involved in improvements of agro-industrial rural development. The selected cases, as proved by an

overview of international circumstances and contexts, cover the same areas of societal demands to the agro-industrial system, which were observed around the world. The key insight of industrial paradigm-based social movements is that current literature on farmers' social movements is too much focused on the struggles of poor farmers and hides the role of rich farmers. The set of case studies on the evolution of national social movements of farmers in Lithuania shows that the social movements representing the interests of large-scale agribusinesses are the most active and have the biggest political power in the industrial era. Each case study demonstrates the efforts of social movements to change political decisions by redistribution of profit, wealth, land and social power on behalf of farmers and other rural populations.

Therefore, more sense in scientific reasoning was found by making a distinction between the industrial paradigm-based and post-industrial paradigm-focused rural social movements, which refer to a rural development paradigm shift by proposing the two scenarios for current rural development strategies, i.e. first, revision; and, second—a fundamental change. Revisionists by propagating the agro-industrial system as the best and the only way forward for human development and by suggesting the discourse of overcoming negative effects, compose a particular branch of rural social movements, concerned with new technological innovations, that help overcome negative outcomes of industrial agrotechnologies. The revisionists advocate the concept of sustainability as a balance among environmental, economic and social needs by socially and conservation-conscious approaches and thus accelerate the belief that the agro-industrial system has no alternatives, accordingly huge share of the public financial resources should be devoted to maintaining the sustainability of industrial agriculture model. Scientific evidence, provided in this monograph proves that many rural social movements support this idea, demanding ever increasing financial assistance for farmers by many supportive measures that should be taken by the governments. However, such approaches to sustainability, focusing largely on efficiency and productivity improvements, as well as 'greening' supply chains and products do not go beyond the industrial paradigm.

Plenty of evidence in the monograph show how the activity of social movements in rural development is changing because of the industrial paradigm shift. Transdisciplinary examination of industrial and post-industrial social movements through the prism of paradigm innovations opens up new ways of understanding the demanded thinking and doing in the times of rural development paradigm shift.

Necessarily, the research of social movements in the context of rural development paradigm shift moves to a further essential stage of the debate, concerning the actors of rural social movements. Overall analysis suggests the existence of a new group of actors that accelerate rural paradigm shift with help of social movements—urban population. The agents of the urban population, such as consumers' movements, back-to-the-land movements, and ecovillages/co-housing movements take another major role in the rural development paradigm shift by broadening, deepening and regrounding the understanding and essence of rural areas and their development. Urban population as actors of change participates in rural development shift in many different ways and thus takes the innovators' role, and there are many cases, they become drivers of post-industrial paradigm focused new rural (hence—being urban agents!) social movements.

Plenty of evidence had been collected around the world, how the change of customer and food/farmer relationships occurs in the light of consumers' movements, back-to-the-land movements, and ecovillages/co-housing, when they start practicing new on-form activities, diversification, servitization, gastronomy and ecological and transformative tourism, healing, amusement, implementation of local food markets. Another huge part of the urban population appeared intending to change the relationships with neighbours, help establish direct relationships with farmers, implement new forms of purchasing variety of product from farmers and rural handcrafters and thus they are radically changing urban–rural relationships in propagating a new form of simple and quick business, based on non-technological innovations, but focused on newly established small-scale urban–rural business models. At the same time, they propagate the value of agroecology, eco-settlements, high quality, and locality of products as a new life quality value, which requires establishing the well-organized short food supply chains and educated consumers. All of this again gives the signals of the already stared rural development paradigm shift, which had been undoubtedly accelerated by new post-industrial mindset-driven social movements. They put into action the threefold complex urban–rural relations instead of previously non-existed or simply interpreted urban–rural connection, namely: (1) satisfying needs for high-quality food and support for sustainable economic development; (2) valorized products with comparative advantage and the transition to agroecology; (3) increased value of knowledge on permaculture and valorized traditional products.

Rural population, namely, small farmers, small agribusinesses and enterprises, as well as agricultural cooperatives also play new roles in the context of the rural development paradigm shift, next to that accomplished by industrial agriculture economic functions, i.e. provision of food and fiber. The rural population as agents of change, next to their direct support to local economic development, guarantees the needs of auto consumption and provides new ways of employment in rural areas, which is specifically relevant in times of crisis when the question of economic resistance to uncertain situations is of top importance. Moreover, the rural population acts as change agents in such important fields as a shift from intensive agricultural industrial technologies to agroecology practices and valorization of traditional products.

Comparing rural and urban population, from a first glance their primary goals might be different, but the final goal is undoubtedly the same—sustainable rural development, rural welfare empowering social, economic and cultural development and rural development which is not possible without non-technological drivers of progress, service-driven business model and especially collaborative and synergistic relationships between both—rural and urban population. The analysis of various stakeholders in social movements and their role in social change proves integrated rural development being a new paradigm, grounded on actively created new practices, which are already mature into specific development trends. The creation and consolidation of new interlinkages between rural and urban actors and between agriculture and society at large prove that the post-industrial rural development paradigm is based on non-technological drivers, service-driven business models and collaborative and synergetic relations that are the signs of the rural paradigm shift, grounded on the mentioned three key paradigm innovations. A part of the population instinctively takes the roles as agents of change and accelerate the raise of particular social movements, to change the state of diminishing but still alive industrial principles in rural areas, thus striving for the new paradigm, which will ensure welfare through collaborative and synergetic joint development.

Moving further, both rural and urban actors, by acting for a rural paradigm change, essentially went through the promoting and restricting circumstances, caused by rural policy directions at international, national and local levels, which also are forming the particular background and shaped the horizons of their activity. Currently, the rural policy focus on neo-endogenous or new internal model of rural development, covering economic, social and environmental dimensions, which steadily refer to the key question of ensured and increased capacity of rural areas by using processes, resources and actions outside their territory in a way that maximizes their benefits. Hence, within the framework of this model, the focus is put on the dynamic interactions between localities and the external political, institutional and economic environment and how these interactions are mediated (Ward et al., 2005). And this proposes that rural development policy cannot exist without appropriate collaboration and networking, to ensure much greater stakeholder involvement and continuous dialogue regarding the governance of this policy. Thus the future rural development policy formation will necessarily involve a very large number of actors in the rural development process at different levels within a variety of socio-economic contexts and having a wide range of needs, priorities, interests and expectations. Hence this kind of networking and collaboration had already been started with the broad geographical and political scale H2020 project SHERPA—Sustainable Hub to Engage into Rural Policies with Actors (https://rural-interface s.eu/), which currently serves the European Commission as policy advice in composing the vision of Rural Europe 2040, elaborated by regional and national acting groups throughout the whole Europe. The given practice of bottom-up networking purely marks a transition towards the new way of thinking regarding the issues and trends of rural development, and, accordingly, it outlines the emergence of a new rural development model based on a new rural development paradigm.

The additional power to rural development policy changes provides the European Green Deal implementation, which is necessarily shaping the activities of social movements in the field. Despite the fact that the European Green Deal is titled as 'the first man on the moon' step for environmental and climate change policy, greater concerns are expressed on its implementation questions. Therefore, not only the top-down approach but also the bottom-up initiatives (e.g. smart specialization, smart villages), which might become drivers of innovation-based practices of particular post-industrial paradigm-focused social movements, are extremely important. Stating the fact, that the history of new social movements demonstrated how climate and environmental action can be done only through a bottom-up approach and supported by the participation of local communities, European Green Deal would fulfill the system with top-to-bottom, long-term mobilization of public and private efforts in favour of a just transition towards a climate-neutral and sustainable economy, leaving no one behind. Thus smart rural communities with their already established and tested social movements are ready to provide many inspiring examples of how to address each of the challenges identified in the European Green Deal at a local level. The European Green Deal will act as a promoting medium for the post-industrial paradigm-focused rural social movement in the context of a rural development paradigm shift, which will mobilize local communities, accelerate collaborative arrangements with local group leaders, researchers and political actors. Indeed, thus the spectrum of top-bottom areas, provided by the European Green Deal, opens up a favourable acting arena with plenty of new possibilities for bottom-up actors to accelerate the rural development paradigm shift with their much promising and already tested eco-innovations. Hence, the spectrum of top-down and bottom-up actors is tremendously wide, and the drivers or the reasons for their participation in various rural activities are diverse. The overviewed European top-down and bottom-up actors demonstrate that the letter (actors), initiatives, organizations and movements focus on many overlapping problems and issues. However, much more important concerning the rural development paradigm shift is the observation that no longer the main focus of both top-down and bottom-up actors is only agricultural issues, as well as exceptionally local areas. Recently European top-down and bottom-up actors play important role in much broader sectors and issues, such as migrants and business, empowerment of women and youth, quality of life of rural residents and indigenous people, food and human rights. And this trend refers to the ties of these movements to post-industrial social movements. In many overviewed cases, their development is characteristic to post-industrial social movements, as they are already changing the identity of rural areas and their perception—so they add to the rural development paradigm shift with their innovations.

Summarizing the systematic analysis of key ideas on rural development through the prism of industrial paradigm shift framework, it was observed that the initiatives of social movements dealing with a generation of radical non-technological innovations for rural development are mainly concentrated on revision and change of industrial mode of relationships. The five case studies from Lithuania demonstrate the characteristics of a new generation of social movements involved in post-industrial rural development, including goals, stakeholders, strategy and organizational structure. Therefore, each case study demonstrates that a new generation of social movements is conceptually focused on the reorganization of current competitive and exploitative relationships to collaborative and synergistic relationships in the context of the growing multifunctional role of the countryside and seeks mutual benefits for both the rural and urban population.

The overall investigations, made throughout this scientific monograph, demonstrate the consistency, expressiveness and impacts of an already ongoing rural development paradigm shift, which is based on paradigm innovations. These innovations had already changed the everyday life of both the rural and urban population, which actively takes part in social movements, depending on their values, beliefs and requirements for the future of rural areas. The paradigm changes are evident, and at least the time and ambitions will judge how quickly the society will accept and adapt post-industrial paradigm innovation recipes, suggested by the new generation of social movements for rural development.

Reference

Ward, N., Atterton, J. H., Kim, T. Y., Lowe, P. D., Phillipson, J., & Thompson, N. (2005). *Universities, the knowledge economy and 'neo-endogenous rural development'* (CRE Discussion Paper). https://a10-vip-wwwprod.ncl.ac.uk/media/wwwnclacuk/centreforruraleconomy/files/discussion-paper-01.pdf. 21 October, 2020.

Index

R. Vilkė et al., *Rural Economic Developments and Social Movements*, Palgrave Advances in Bioeconomy: Economics and Policies, https://doi.org/10.1007/978-3-030-71983-8

D

The manufacturer's authorised representative in the EU is Springer Nature Customer Service Centre GmbH, Europaplatz 3, 69115 Heidelberg, Germany. If you have any concerns regarding our products, please contact ProductSafety@springernature.com

Printed and bound by CPI Group (UK) Ltd, Croydon, CR0 4YY

30/09/2026

02225457-0008